T0222649

Playing with Infinity

This is a book about infinity — specifically the infinity of numbers and sequences. Amazing properties arise, for instance, some kinds of infinity are argued to be greater than others. Along the way the author will demonstrate how infinity can be made to create beautiful 'art', guided by the development of underlying mathematics. This book will provide a fascinating read for anyone interested in number theory, infinity, math art, and/or generative art, and could be used a valuable supplement to any course on these topics.

Features

- Beautiful examples of generative art
- Accessible to anyone with a reasonable high school level of mathematics
- Full of challenges and puzzles to engage readers.

Hans Zantema was born in 1956 in The Netherlands. He studied mathematics and received his PhD in pure mathematics in 1983. After a few years in industry he returned to university, in computer science. Apart from his position as an associate professor in computer science at Eindhoven University of Technology, from 2007 until his retirement in 2022 he was part time full professor at Radboud University in Nijmegen. His professional interest is mainly in mathematical reasoning, in particular applied to computation and automated reasoning. His hobbies include solving and designing logical puzzles.

AK Peters/CRC Recreational Mathematics Series

Series Editors:
Robert Fathauer, Snezana Lawrence, Jun Mitani, Colm Mulcahy,
Peter Winkler, and Carolyn Yackel

The Baseball Mysteries
Challenging Puzzles for Logical Detectives
Jerry Butters and Jim Henle

Mathematical Conundrums
Barry R. Clarke

Lateral Solutions to Mathematical Problems
Des MacHale

Basic Gambling Mathematics
The Numbers Behind the Neon, Second Edition
Mark Bollman

Design Techniques for Origami Tessellations
Yohei Yamamoto and Jun Mitani

Mathematicians Playing Games
Jon-Lark Kim

Electronic String Art
Rhythmic Mathematics
Steve Erfle

Playing with Infinity
Turtles, Patterns, and Pictures
Hans Zantema

For more information about this series, please visit: https://www.routledge.com/AK-PetersCRC-Recreational-Mathematics-Series/book-series/RECMATH

Playing with Infinity

Turtles, Patterns, and Pictures

Hans Zantema

CRC Press
Taylor & Francis Group
Boca Raton London New York

CRC Press is an imprint of the
Taylor & Francis Group, an **informa** business

AN A K PETERS BOOK

Front cover image: Victor Habbick Visions/Science Photo Library

First edition published 2024
by CRC Press
2385 NW Executive Center Drive, Suite 320, Boca Raton FL 33431

and by CRC Press
4 Park Square, Milton Park, Abingdon, Oxon, OX14 4RN

CRC Press is an imprint of Taylor & Francis Group, LLC

Library of Congress Cataloging-in-Publication Data
Names: Zantema, Hans, author.
Title: Playing with infinity : turtles, patterns, and pictures / Hans Zantema.
Description: First edition. | Boca Raton : AK Peters/CRC Press, 2024. |
Series: AK Peters/CRC recreational mathematics series |
Includes bibliographical references and index.
Identifiers: LCCN 2023048349 (print) | LCCN 2023048350 (ebook) |
ISBN 9781032738000 (hbk) | ISBN 9781032706108 (pbk) | ISBN 9781003466000 (ebk)
Subjects: LCSH: Infinite–Popular works. | Mathematical recreations.
Classification: LCC QA9 .Z36 2024 (print) | LCC QA9 (ebook) |
DDC 515/.24–dc23/eng/20231122
LC record available at https://lccn.loc.gov/2023048349
LC ebook record available at https://lccn.loc.gov/2023048350

ISBN: 978-1-032-73800-0 (hbk)
ISBN: 978-1-032-70610-8 (pbk)
ISBN: 978-1-003-46600-0 (ebk)

DOI: 10.1201/9781003466000

Typeset in Optima
by codeMantra

Contents

What is this book about?

What is the largest number?
How many numbers are there?
Can we store all those numbers in a computer?
Numbers that we will never encounter, do they actually exist?

Many questions that we may ask ourselves. To start with the first question: every number can be increased by simply adding one to it. So no number is the largest number, and the largest number does not exist. And that also answers the second question: the number of numbers that exist is greater than any number you can imagine.

Infinitely many

There is no other option than to call that number of numbers *infinity* : a quantity greater than any finite number. But what exactly is infinity?

Everything we store in a computer is ultimately stored as a string of zeros and ones. This applies not only to all documents, including this book, but also for all pictures, photos, music files and movies. Over the years, the available memory of a computer has improved considerably. While being far beyond feasibility 40 years ago, now you easily store a complete movie in high resolution and high sound quality on a computer. But no matter how this will improve in the forthcoming years, it will always remain finite. Where memory used to be measured in megabytes, now mostly in gigabytes, and the total memory runs into terabytes, it will no go beyond finiteness. Whatever smart techniques are developed, more than all the atoms in the universe will never become available to store files. And even though we don't know exactly how many atoms there

DOI: 10.1201/9781003466000-1

are in the universe, it is very unlikely that there are infinitely many. And even if there would be infinitely many atoms, only a finite part of them would be reachable in reasonable time due to the limits on speed of light.

A file containing all natural numbers :

$$0, 1, 2, 3, 4, 5, 6, 7, 8, 9, 10, 11, 12, 13, 14, 15, 16, 17, \ldots$$

going on indefinitely so that every natural number occurs in it will be infinite. So you will never be able to save this, not on all current computers of the world together, nor on all the world's future computers.

This intriguing concept of infinity will be the main topic of this book. First we will look at different kinds of numbers, all of which are infinite, and then we will argue that one kind of infinity is greater than the other.

Infinite sequences and turtle figures

After considering numbers and flavors of infinity, we will focus on infinite *sequences*. In their simplest form, they only consist of zeros and ones, as the logical way to store infinite things, just like finite series of zeros and ones are the logical way to store finite things on a computer.

In particular, we will focus on a simple way to make such sequences visible in a picture using *turtle figures*. This will be discussed in detail later, now we restrict to giving an impression of what's going on. To both symbols 0 and 1, we associate some angle. A robot, often referred to as a *turtle* for historical reasons, reads the symbols of the sequence one by one, starting at a particular point in the plane and at a particular running direction. First, the turtle reads the first symbol in the sequence: if this is a 0 the running direction is turned over the angle associated to 0, and if it is a 1 it is turned over the angle associated to 1. And then a *step* is done, of the length of a fixed unit. This will be repeated forever: the next symbol from the row is read, the running direction is turned according to the corresponding angle and another step is done, and so on. The turtle figure is then the route that is traveled. We give a simple example to illustrate the idea.

Let's look at the sequence consisting only of zeros, and where the angle associated with 0 is equal to $160°$. The angle associated to 1 then does not matter because it does not occur in the sequence anyway. So the turtle figure is obtained by alternating endlessly rotating the direction through 160° and doing a unit step. That means that in every next step, a sharp angle of 20° is made. After nine steps, the following figure has been obtained:

and after these nine steps, both the location and the running direction is exactly the same as in the beginning. So by continuing every nine steps, the given figure will be drawn over again, in which only segments are drawn that were drawn before. So the given star figure is not only the turtle figure for the first nine steps but also the turtle figure for the infinite sequence.

Morphic sequences and symmetry

This simple star figure is not yet complicated and surprising, but a lot more exciting is the following. Consider the sequence

$$\mathbf{t} = t_0 t_1 t_2 \cdots = 0110100110010110 \cdots$$

which is defined by $t_0 = 0$, $t_{2i} = t_i$ and $t_{2i+1} = 1 - t_i$ for each $i = 0, 1, 2, 3, \ldots$ For the moment don't worry on such a technical formula, later in the book this sequence and its definition will be explained in more detail. This sequence is called the Thue-Morse sequence. We now show the turtle figure we get for this sequence if the angle for 0 equals $-146.25°$ and the angle for 1 equals

4.74609375°. For the time being, these numbers look weird, but these will be explained too.

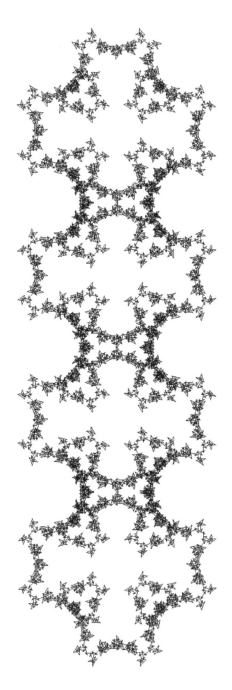

Here we reduced the step size to about a millimeter in order to get a figure fitting on the page. As in the first example, after a number of steps (in this case not 9 but 16,384) the turtle is back where it started, both in position and direction, and everything after this initial part will consist of exactly the same steps that have been done before, although the remaining part of the row is not at all equal to the whole row. Why that's the case is at exactly these angles, and how we came up with the idea of choosing these weird angles is something we'll see later as a result of a careful analysis. That the picture has all kinds of symmetry can also be argued, but how the resulting picture eventually looks like, being the result of a computer program of only a few lines, remains a surprise. As said: much more about this later; at this point, we only remark that if we replace the angles by arbitrary other angles, almost always the resulting turtle figure is a completely chaotic picture in which the starting point will be never met again.

The Thue-Morse sequence we just saw has another special feature. If we replace every 0 by 01 in this sequence, and at the same time every 1 by 10, the result is again exactly the Thue-Morse sequence itself. And the Thue-Morse sequence is the only infinite sequence that starts with a 0 and has this property. This idea will be generalized: if we have a finite sequence $f(0)$ of zeros and ones starting with 0, and choosing a finite sequence $f(1)$ of zeros and ones, then we can create an infinite sequence $f^\infty(0)$ that starts with 0 and has the property that if in that infinite sequence very 0 and 1 are replaced simultaneously by $f(0)$ and $f(1)$, exactly the same row is obtained. Such sequences are called *morphic*. So the Thue-Morse sequence is the morphic sequence obtained by choosing $f(0) = 01$ and $f(1) = 10$. We have already seen an example of a turtle figure of the Thue-Morse sequence that returns to the starting point after finitely many steps, and after that finite part only line segments will be drawn that were drawn before. So by drawing the turtle figure of this finite initial part, we have the turtle figure of the full infinite sequence. We will also see such finite turtle figures for other morphic sequences, and by choosing the right angles we often get nice symmetrical patterns. Further details will be discussed later, here we only show the following example with a 17-fold symmetry:

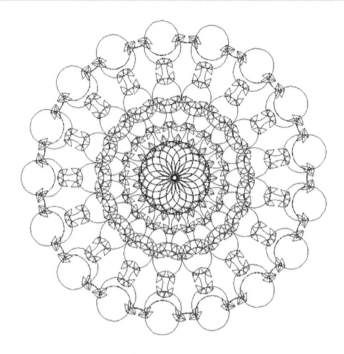

Fractal turtle figures

Not all interesting turtle figures are finite. We will also see *fractal* turtle figures. An infinite collection of points in the plane is called *fractal* if it has the wonderful property that if you scale up the distances between all points by a conveniently chosen fixed factor, you only get points that were already in the original collection. We will show that under certain conditions the set of points of a turtle figure of a morphic sequence is fractal. More precisely: the set of the end points of all drawn line segments in the turtle figure is then a fractal set. Again, further details will be discussed in detail later, here we only give an example of a part of a fractal turtle figure obtained accordingly:

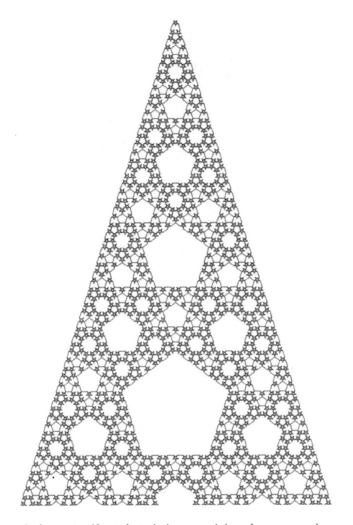

The turtle figure itself is infinitely large and therefore cannot be presented completely. Only a part is shown here: the full turtle figure starts at the top, but continues infinitely downwards and also to the left and right.

Turtle figures of morphic sequences are full of all kinds of patterns. An example of this is the following finite part of an infinite turtle figure.

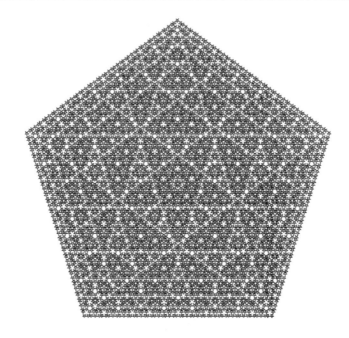

Mathematical challenges

In any mathematical framework, including the theory of turtle figures, morphic sequences and fractals, it is important to define concepts carefully, and to investigate which properties hold. Without such a systematic analysis, the pictures just shown would never have existed, and this book could not have been written.

When thinking of math, many people think of the math they used to have in school, about how to do calculations, and the tricks that should be mastered in order to pass a test. However, setting up a mathematical framework is a lot more exciting: you give definitions yourself and then find out which properties hold. For these properties, you want to be sure that they hold. The only way to be sure is to formulate the property very precisely as a theorem, and then check its validity by giving a proof, i.e. a logical reasoning proving that the statement is true. From that moment on, you don't have to worry about that anymore, and you can always use that statement because you already know for sure that it is true. That's the basis of the framework, and the framework may be extended by adding and proving new theorems, in which existing theorems may be used. A great part of mathematics, theoretical physics, theoretical computer science and logic can be seen as an accumulation of such a framework. People working in these areas are very familiar with all the basic propositions that apply there, being their tools to extend the framework and arrive at new results.

These frameworks can be very extensive, and a great deal of knowledge of the underlying layers is required in order to understand what is happening at the top. For the ultimate results, it may be hard to reach a wider audience than just the specialists in such a field, to clarify what it is all about. In popular science media, and science supplements of newspapers, sometimes attempts are made. Usually the emphasis is then on the results, for example about exotic particles such as neutrinos, unexpected behaviour in galaxies very far away, or pictures of black holes, or the discovery of a new largest known prime number. What I personally observe in such explanations is that they are only about results that are hard to understand for a layman, and hardly give any feeling about how the achievements were obtained.

This is in a sharp contrast to doing science yourself: that is an exciting journey of discovery in which you ask yourself all kinds of questions and try to answer them. Some may have a simple solution; for most of them, it is impossible to get even a snippet of an answer, but occasionally a surprising answer to a question can be given, which can then be further built on. Very important in such a process is to ask yourself the right questions, and the actual elaboration can be a huge puzzle. When succeeding after a lot of effort, this may give a lot of satisfaction. In the final publications, the emphasis will be on the results obtained, and hardly anything will be visible of all struggle and many failed attempts. But in order to reach these results, all struggle may be of great value.

I am happy to have experienced all of this myself extensively. The study of pure mathematics in the late 1970s in Groningen was already great, and after that I got the opportunity to do a PhD research in Amsterdam in the field of algebraïc number theory. On the one hand, it was wonderful, but on the other hand, it was so specialized that I sometimes thought: what is the goal of all of this? Then I switched to work in a company for a few years. For sure this was an interesting experience, but it also showed that I was too much a scientist for being happy in such an environment for a longer time. So I really appreciated that in 1987 I got the opportunity to return to science and switched to computer science at Utrecht University.

After looking around in several topics, I found my own way there. My focus was on all sorts of topics that originate from computer science but are very mathematically oriented, and over the years, I succeeded in publishing a lot. In 2000, I switched to the Technical University in Eindhoven, again in computer science, a position that was extended by a part-time full professorship at Radboud University in Nijmegen in 2007. And all these years, I really enjoyed doing both research and teaching: a golden combination. What I already outlined above applies to my research: many of the techniques and results that I developed require too much specific knowledge to be able to explain to a wide audience. Clearly, if I would have done only simple things, Radboud University would not have appointed me as a full professor.

But on the other hand, in those forty years of research, I came across surprising observations a number of times that do not require a lot of specific prior knowledge. For me it is a challenge to share this with a wider audience than just direct colleagues. And that's one of the main goals of this book: to join the reader in a scientific adventure leading to some remarkable discoveries. And then in such a way that it is not only about the results but also about the search process to reach them, and all the puzzles encountered along the way. That is what scientific research in such a direction makes so much fun. On the way during this book, we will meet several snippets of theory, some of which have been widely known for hundreds of years, while others showed up only recently in my own research.

The journey of discovery that I have in mind for the reader, certainly will not only consist of passively reading the explanations I wrote down. It is mainly about puzzling and thinking yourself, being the core of doing research. To that end, the reader is occasionally invited to solve some problem. More precisely, this will be done by finishing each chapter by a *challenge*. Some of them will be really tough, while some others are more doable, just like the challenges you may encounter in doing research. In fact, some of those challenges have actually shown up in my research, and the solutions appeared in my publications. For none of them, the solution is obtained: you really have to work for it. And these challenges all have in common that their solutions will be given in the course of the book, so if you fail after heavy attempts (which you should not be ashamed of at all) you may just continue reading, or maybe flip back, and for sure you will find a solution with explanations somewhere in this book.

Who is this book for, how is it organized and how to read it?

Who is this book for? This book contains a lot of turtle figures of infinite sequences, and if you like these pictures, this book is for you. But this is only part of the story: this book also tries to explain how these pictures are constructed, and why they have their remarkable properties. For understanding this, some mathematical framework is built, and the pictures appear as an interplay between this mathematical framework and experimenting with computer programs generating the pictures. So for full understanding, some knowledge and feeling for mathematics will be required, but this will never go beyond high school level. This even holds for the solutions of the challenges: finding solutions of the challenges yourself may be very hard, that's why are called challenges, but for finally understanding the solutions as they are given in this book, no deep mathematics is required. So, in general, we may say that this book is intended for all people that like logical puzzles and mathematical reasoning and have at least a high school background in these.

The main topic of this book is turtle figures of infinite sequences. But details of infinite sequences do not show up earlier than in Chapter 5, at one quarter of the book, and turtle figures show up even later, in Chapter 6. In the first chapters, some general theory on infinity is presented: basics on the set of natural numbers in Chapter 2, basics on integer, rational, real and complex numbers in Chapter 3, and some basics on different kinds of infinity in Chapter 4. The reason for this is not only that it makes sense to start by these basics of the notion of infinity. Some of these notions also play a role in observations we make on infinite sequences and their turtle figures. For instance, finiteness of turtle figures of periodic sequences shows up to be closely related to choosing angles being rational numbers. On the other hand, if you ar mainly interested in these turtle figures, it should be possible to skip this first part and start in Chapter 5 and then skip references to the earlier chapters.

After some basics of infinite sequences in Chapter 5, and the notion of turtle figures in Chapter 6, this book continues by discussing how to program turtle figures in Chapter 7. In Chapter 8, *morphic sequences* are introduced, being the main class of sequences of which the turtle figures are investigated. One particular morphic sequence is the Thue-Morse sequence, being the topic of Chapter 9. Here we still focus on finite turtle figures; more finite turtle figures having nice symmetry are discussed in

Chapter 10. In Chapter 11, *fractal* turtle curves are presented. A particular fractal turtle figure related to the *Koch curve* is discussed in Chapter 12. Chapter 13 investigates the simples possible morphic sequences and their turtle figures. Finally, Chapter 14 looks back on the whole book, and presents some other flavors of turtle curves and some other ways to generate nice pictures by small programs. And all these chapters are concluded by challenges and are mixed up by solutions of these challenges.

How to read this book? When you read a novel, or an article in a newspaper or magazine, you start at the beginning, and read until you've read it all, and then you're done. Reading this book may be very different, just like I read a scientific book or article myself: I read a bit, and what I read triggers my mind: I try to understand something or think about something, and make notes on a piece of paper. For me a piece of paper is an indispensable tool for reading this kind of text. Occasionally I fall back on reading the text. It hardly ever happens that I read such a book or article completely from start to finish. Typically, some pieces of text I take for granted without grasping all details. On the other hand, going through a difficult proof and understand it may take a lot of time. Spending a full day reading and understanding one page is no exception at all in the scientific process, and often may be experienced as a well-spent day. Hopefully, the challenges in this book will have that effect on some readers. How you read this book can vary greatly from person to person. Some readers may want to understand, and some of the pictures shown without going through all the evidence, or maybe even just looking at those pictures: well, let everyone feel free to skip whatever you want if it's difficult or tough or gets boring. But someone else might be interested in that, and for them I try to give the story as complete as possible. Providing evidence also plays an important role in the whole: by providing evidence you find out exactly which conditions are essential, and those insights in turn guide the search for examples. The series of turtle figures in this book could only appear together with the careful development of the accompanying theory, and that is why that theory is such a crucial part of this book.

As this introductory chapter is now close to finish, it's time for the first challenge. And that is a very special one, really very difficult. So this definitely does not follow the basic didactic principle to start simple. It is a challenge that emerged in my own research not so long ago. To be precise:

that was in the spring of 2021, triggered by an open problem presented by a colleague.

Challenge: the paint pot problem

Imagine you have a finite sequence of paint pots. Every pot is either empty or filled with paint of some color. Now you can adjust this sequence according to the following rules:

- You may swap two adjacent pots that are both filled with paint.

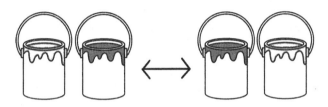

- If both neighbors of a filled paint pot are empty, you may distribute the paint over the two empty neighbor pots, after which the middle one is empty and both neighbors are filled with the same color paint. This step may also be reversed: if both neighboring pots of an empty pot contain the same color of paint, you may empty both neighboring pots in the middle pot, after which both neighboring pots are empty. Here we ignore possible overflow.

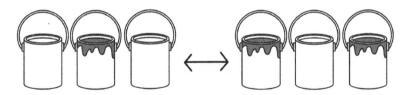

Challenge:

Is it possible to start with a sequence of paint pots in which the first four pots contain paint in four different colors, and by only applying steps of the above type end up in a sequence of which the first pot is empty?

This is all: a precisely formulated question to which the answer is 'yes' or 'no', and you are asked to find out what the correct answer is. Of course you want to be sure that the answer is correct. If the answer is 'yes', you should show this by giving a particular sequence of paint pots of which the first four are filled with different colors of paint, and then show all the steps ending in a sequence of which the first pot is empty. If the answer is 'no', then you want to provide a formal argument showing so. As promised, the answer and the full argument will be given later in this book.

For the time being, we leave open whether the answer is 'yes' or 'no', we only suggest a notation to be used. Let's write *a* for an empty pot, and *b*, *c*, *d* and *e* for pots with the different colors of the first four pots. Perhaps there are more colors, but they will turn out to play no crucial role, so we'll leave that out for now. Now we describe a sequence of paint pots by a sequence of symbols *a*, *b*, *c*, *d*, *e*. The first rule states that you always may swap two neighbors that are not *a*. The second rule states that the pattern *aba* may be replaced by *bab* and vice versa, and similarly if here *b* is replaced by *c*, *d* or *e*. The question of the paint pot problem then is whether it is possible to start with a sequence that starts with *bcde*, just doing these kinds of steps can end up with a sequence that starts with *a*. The nice thing is that you don't have to worry about the interpretation anymore of the paint pots, and overflow issues and the accompanying mess. Using this notation, the paint pot problem boils down to this question only about the symbols *a*, *b*, *c*, *d*, *e*.

In the paint pot problem, you now have to start with a sequence that starts with *bcde*, for example *bcdeaeda*, and with that, you can do the following steps:

$$bcd\underline{ea}eda \rightarrow bcdae\underline{ada} \rightarrow bcda\underline{ed}ad \rightarrow bc\underline{da}dead \rightarrow bcadaead.$$

The part to which a rule is applied is always underlined, and in this example the third pot in the sequence became empty. The question now is whether it is also possible with a possibly much longer sequence starting with the first four letters also being *bcde*, and eventually ending up with a row that starts with *a*.

Good luck!

2

Numbers of the simplest kind

In Chapter 1, we already mentioned *natural numbers*:

$$0, 1, 2, 3, 4, 5, 6, 7, 8, 9, 10, 11, 12, 13, 14, 15, 16, 17, \ldots$$

The set of natural numbers can be considered as the simplest infinite set, so this justifies a closer look at these natural numbers in this book on infinity.

Natural numbers

How to define or describe natural numbera? The easiest way to do that is using the following rules:

- 0 is a natural number, and
- if n is a natural number, then its successor, successor, $s(n)$ is also a natural number.

Now a natural number is by definition something that can be made by applying these rules finitely often, and all results of this process are assumed to be distinct. So the natural numbers are

$$0, s(0), s(s(0)), s(s(s(0))), s(s(s(s(0)))), \ldots$$

Introducing $1, 2, 3, 4, \ldots$ as a shorthand notation for, respectively, all of these numbers, exactly gives the natural numbers as we know them. These rules are the core of Peano's axioms, as described by the Italian Giuseppe Peano (1858–1932). More precisely, among the five axioms two exactly coincide with our two rules, two describe that the resulting natural numbers are all different, and the *induction axiom* describes that all natural numbers can

DOI: 10.1201/9781003466000-2

be reached in this way. The original set of Peano's axioms also covers some general logical principles about how to deal with equality.

Some people argue that the natural numbers should not start at 0 but at 1, this may be a topic of endless discussion. In fact, in Peano's original axioms, 1 is the starting point, not 0. However, for many issues, it does not matter at all. Currently, the most common is to start with 0, and that also has several advantages. So, in this book, the natural numbers will always start at 0.

The natural numbers can be seen as a kind of starting point for the whole framework of mathematics.

In Chapter 3, we will see how to build richer number systems based on the natural numbers as a starting point: integer numbers begin natural numbers extended by negative numbers, rational numbers being integer number extended by fractions, real numbers as rational numbers extended by limits, and even complex numbers. Here we mention a statement about mathematics that is attributed to the German Leopold Kronecker (1823–1891): *God created the integers, all the rest is human work.*

But first, in this chapter, we focus on how to deal with natural numbers: how to use them for definitions, how to prove properties, what operations appear on natural numbers, and what properties hold for these operations. These are insights that are known for hundreds of years old. They are included in this book for several reasons: as an overview of important properties of the infinitely many numbers of the simplest kind but also as an example of how to set up a theory and how to reason with such a theory.

Induction

By definition, a natural number is something that can be obtained by starting at 0, and then apply successor a finite number of steps. This is Peano's axiom of induction. Now we will use this definition as a way to define something new: first define it for 0 and then define it for the successor of n, i.e. for $n+1$, assuming you already know what comes out for n. Such a definition is called an *inductive definition*. Later in this section and in forthcoming sections, we will see several examples.

But we can also use the same principle to give a *proof*, that is, a solid argument that a statement holds. If we want to prove that a property about natural numbers always holds, for every natural number, we can do so by proving the following:

- Show that the property holds for 0, which we call the *basis* of the induction, and
- Assume that the property holds for a number *n* (we call this the *induction hypothesis*), then we show that the property also holds for $s(n) = n + 1$. We call this second part the *induction step*.

If both parts hold, then we know from the first part that the property holds for 0. But then due to the second part, the induction step, we know that it also holds for $1 = s(0)$. But since it holds for $s(0)$, again by using by the induction step, we know that it also holds for $2 = s(s(0))$. And so on for $3, 4, 5, \ldots$ Since every natural number can be obtained in this way in a finite number of steps (that was Peano's axiom of induction), then we may indeed conclude that the property holds for every natural number.

A proof of a property about natural numbers given in this way is called an *induction proof.*

Let's give an example. Suppose we want to know what is the result of adding the numbers from 1 to 100. Of course it can be done by doing 99 additions separately, but it can also be done smarter. Here 100 is just a number, let's see what comes out when we add the numbers from 1 to *n*, and let's call the result $T(n)$. These numbers are called *triangular numbers*, that's why we use T for it. The reason they are called triangular is that they coincide with the number of circles in the triangle of the following kind:

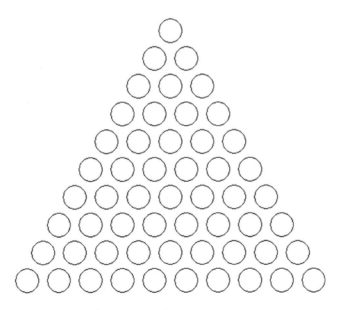

In the top row, we see one circle, in the second row two circles, and so on until the tenth row with ten circles. The total number of circles is therefore $T(10)$. Now we will give a general formula for $T(n)$, that we will prove to be correct by induction. First we give an inductive definition for $T(n)$. The number $T(0)$ is the number of circles in a triangle of size 0, which is 0. If $T(n)$ is the number of circles in a triangle of size n, then we get a triangle of size $n+1$ by adding a row of $n+1$ circles, which yields a total of $T(n)+n+1$ circles. So this yields the inductive definition:

- $T(0) = 0$, and
- $T(n+1) = T(n) + n + 1$.

Indeed this has the structure of an inductive definition, so $T(n)$ is defined for any natural number n in this way. Now we may ignore the pictures with triangles: it is just a matter of finding a formula for $T(n)$. By choosing small numbers we find $T(1) = 1$, $T(2) = 3$, $T(3) = 6$, $T(4) = 10$, $T(5) = 15$. These all satisfy $T(n) = \frac{n(n+1)}{2}$, so we expect this to be true for every n. And that is exactly the claim that we will now prove by induction. First the basis: for $n = 0$ we have: $T(0) = 0$ by definition, and we have $\frac{0(0+1)}{2} = 0$, so indeed $T(0) = \frac{0(0+1)}{2}$, by which the basis of the induction proof holds. Next the induction step: we assume that $T(n) = \frac{n(n+1)}{2}$ for some n, and now we have to prove that $T(n+1) = \frac{(n+1)(n+1+1)}{2}$. According to the definition, we know $T(n+1) = T(n) + n + 1$, and we already know that $T(n) = \frac{n(n+1)}{2}$, so we have $T(n+1) = \frac{n(n+1)}{2} + n + 1$. Using a tiny bit of high school math, we get:

$$
\begin{aligned}
T(n+1) &= \frac{n(n+1)}{2} + n + 1 \\
&= \frac{n(n+1) + 2(n+1)}{2} \\
&= \frac{(n+1)(n+1+1)}{2}.
\end{aligned}
$$

This concludes the induction step, and we have proved by induction that for every natural number n we have $T(n) = \frac{n(n+1)}{2}$. So adding the numbers from 1 to 100 yields $T(100) = \frac{100*101}{2} = 5050$, much easier to compute than adding all 100 numbers.

For this particular example, we now give another argument showing that $T(n) = \frac{n(n+1)}{2}$. Consider the next picture for $n = 10$, but the argument works for any n.

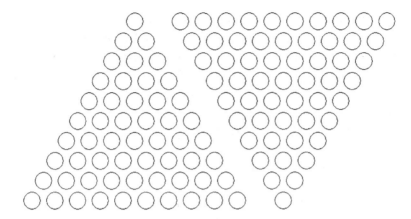

We count the number of circles that we see here in two ways. On the one hand, we see two triangles with $T(n)$ circles each, clearly the fact that one of the triangles is shown upside down does not change the number of circles. So the number of circles is $2T(n)$. But we can also see this as a skewed rectangle of n rows each consisting of $n+1$ circles. So that yields $n(n+1)$. So this shows $2T(n) = n(n+1)$, hence $T(n) = \frac{n(n+1)}{2}$. For our story on induction, this is just a side remark. Induction is an indispensable tool in many branches of mathematics and computer science, and in most cases, it is not possible to find an argument, as here with the formula for triangular numbers, that does not require induction.

Almost everyone who studies mathematics or computer science or something similar will see dozens of this kind of induction proofs in their first year. However, for the moment, we restrict to this single example.

We do note that this formula $T(n) = \frac{n(n+1)}{2}$ is not just an example, but can be used in many ways, for example when solving one of the turtle figure challenges that we will encounter later in this book.

Strong induction

Often it is convenient to use a variant of induction called *strong induction* to prove that a certain property $P(n)$ holds for all natural numbers n. This principle looks like this:

> If we can prove the property $P(n)$ for every n under the assumption that $P(i)$ holds for all $i < n$, then $P(n)$ holds for every natural number n.

In 'normal' induction, we also want to prove a property $P(n)$ for every natural number n, and then first we have to prove $P(0)$, and assuming the induction hypothesis $P(n)$ we have to prove $P(n+1)$. So in proving P for the new value $n+1$, the induction hypothesis $P(n)$ only deals with the direct predecessor of $n+1$. In contrast, in strong induction when proving P for a new value n, we may assume that P holds for all predecessors of n, not only the direct predecessor. This is why it is called *strong* induction. Moreover, as 0 has no predecessors, the property $P(0)$ has to be proved without having any assumption, but in the formulation of strong induction, it is not necessary to formulate this separately.

A surprising observation is that the correctness of this new principle may be proved by 'normal' induction, as we will see now. We prove by 'normal' induction that $Q(n)$ holds for all n, where $Q(n)$ is the statement that for all natural numbers $i < n$ the property $P(i)$ holds. The property $Q(0)$ is true for a very trivial reason because there is no natural number $i < 0$. Next assume that $Q(n)$ holds, then we prove $Q(n+1)$ as follows. We have to prove that $P(i)$ holds for all $i < n+1$. If $i < n$ this follows from the assumption $Q(n)$. And if $i = n$ this follows from the fact that we can prove $P(n)$ under the assumption that $P(i)$ holds for all $i < n$. This proves by 'normal' induction that $Q(n)$ holds for all natural numbers n. But then $P(n)$ also holds for all natural numbers n, since $P(n)$ follows from $Q(n+1)$. This proves the correctness of the principle of strong induction.

The assumption that $P(i)$ holds for all $i < n$ is also called the *induction hypothesis*. Later in this book we will use this principle of strong induction a number of times.

Sometimes this principle of strong induction is used in a hidden way. To prove that $P(n)$ holds for every natural number n, then a proof by contradiction is presented: assume that $P(n)$ does not hold for some n. Then take the smallest number n for which $P(n)$ does not hold. The fact that this smallest natural number exists is a basic property of the natural numbers, closely related to the principle of strong induction. Then the fact that n is the smallest number for which $P(n)$ does not hold implies that $P(i)$ holds for all $i < n$. So using this coincides with using the induction hypothesis in strong induction.

Let's go back to our general presentation of natural numbers. In the example of triangular numbers, we already used addition and multiplication. But what exactly is addition and multiplication? Next we will see how to define addition and multiplication using only our definition of natural numbers as a starting point.

Addition

The first thing you do when learning math at school is *addition*. If you have some number of objects, and you add another number of the same objects, how many objects do you have in total? This is a very natural operation where you enter two numbers n, m and a new number $n + m$ comes out. How can we define this $+$ operation on two natural numbers? We will now do this by an *inductive definition*. Suppose we have an arbitrary m, and now we want to define $n + m$ for every natural number. The first case is $n = 0$, and we need to define what $0 + m$ is. If I have m objects, and add 0 objects, that's nothing at all, then I still have m objects. So the first piece of the definition looks like this:

$$0 + m = m.$$

Next, assuming we already know what $n + m$ is, we want to define what $s(n) + m$ is. Note the parentheses: this is $+$ applied on $s(n)$ and m. Successor means you add one, and for the result it doesn't matter if you add that one to the n first, or if you add the one to the result $n + m$. That's why we define

$$s(n) + m = s(n + m).$$

Nothing exciting seems to be happening, but we observe that by defining it like this, we have a precise inductive definition of the addition of two natural numbers. The entire $+$ operator is completely defined by just the two rules $0 + m = m$ and $s(n) + m = s(n + m)$ that we may use for all natural numbers n, m.

What properties does this addition have? Surprisingly, using this inductive definition, we may prove many kinds of known and desired properties by induction. Let's give an example of this: for every natural number n it holds $n + 0 = n$. You can say: of course it is: if I put 0 at n, it will remain n. From the idea like the teacher at school used to add, that sounds logical. But can we now also prove that with just the definition of addition as we just gave? That means that the only property we know about addition, and therefore the only we're allowed to use, is that $0 + m$ equals m and that for every n $s(n) + m$ equals $s(m + n)$, all this for every m. Now we want to prove that for every n holds: $n + 0 = n$. We do this by an induction proof.

First we need to show the basis, that is $0 + 0 = 0$. Well, it does, because we are allowed to use that $0 + m = m$ for every m, and if we do that for $m = 0$ then it yields $0 + 0 \stackrel{*}{=} 0$, exactly the basis we had to prove. Next, we assume that $n + 0 = n$ (the induction hypothesis), and now we have to prove that $s(n) + 0 = s(n)$. But because of $s(n) + m = s(m + n)$ for $m = 0$ we already know that $s(n) + 0 = s(n + 0)$, and because of $n + 0 = n$ (the induction hypothesis) we now conclude $s(n) + 0 = s(n + 0) = s(n)$, which proves the induction step. Implicitly we used an equality property here: from $s(n) + 0 = s(n + 0)$ and $s(n + 0) = s(n)$ we concluded $s(n) + 0 = s(n)$. To be very precise, we used one of Peano's axiomas that we left implicit here.

In this way, we have proved by induction that $n + 0 = n$ for every natural number n.

Similarly, you can prove that addition *is associative*: for every n, m, k holds $(n + m) + k = n + (m + k)$. Slightly trickier is proving that addition is *commutative*: for every n, m it holds $n + m = m + n$. Here the basis requires that $0 + n = n + 0$, and since we just proved that $n + 0 = n$ and by definition $0 + n = n$, that is correct. But for the induction step you need $s(n) + m = n + s(m)$, which you can first prove by induction separately. You can do this yourself, 'by hand' as it is sometimes called, although your brains will also participate. But you can also use computer search for this kind of evidence, and in the properties mentioned here this can be done automatically. That's not very complicated, in fact I once wrote a program doing so myself, not being sophisticated at all. The program then first tries to prove the new property using the given equations directly. If that fails, the program tries to prove the new property by induction. If that fails too, then it is investigated which extra properties are needed for a proof, and then the program tries to prove these extra properties by induction. Many properties can be proven fully automatically this way. Sometimes if it doesn't work out, you may continue by composing a suitable auxiliary property yourself. It is often the case that such auxiliary property can be proved automatically, and then with that extra auxiliary property added as a rule, the original intended property can proved automatically. This framework can be applied to the equations that define the natural numbers, but also to completely different equations. This branch is called *inductive theorem proving*. Although surprising theorems can be proved fully automatically in this way, for most known properties in natural numbers, like properties on prime numbers as we will discuss later, proving them fully automatically is far beyond the power of inductive theorem proving.

Multiplication

Back in school, the topic of *subtraction* came after addition. But subtraction is nothing new: $m - n$ is the number k for which holds $k + n = m$. Moreover, you can only calculate $m - n$ as a natural number if $m \geq n$, so not for all m, n, and that brings additional problems that we don't want to worry about here. In Chapter 3, we will consider subtraction in the integers.

Now let's do the next real step: *multiplication*. If there are n rows with m objects each, then there are $n \times m$ objects in total. Can we define this $n \times m$ inductively too, just like we did for the addition? Yes. That means we have to do two things: we have to define $0 \times m$, and assuming we know $n \times m$ for some n, we have to define $s(n) \times m$. And both are not that hard: with 0 rows of objects there are 0 objects in total. And $s(n)$ rows each containing m objects are obtained by taking n rows each containing m objects, and then add one more row with m objects, resulting in $(n \times m) + m$ objects. In summary, using the operation $+$ we already had, the inductive definition for multiplication reads:

$$0 \times m = 0, \quad \text{and} \quad s(n) \times m = (n \times m) + m,$$

and all this for every natural number n and every natural number m. Using this inductive definition, multiplication is defined completely.

Multiplication, together with addition, has a lot of nice properties. To name a few, for all natural numbers n, m, k we have:

$$n \times 0 = 0,$$

$$n \times s(0) = n,$$

$$n \times m = m \times n,$$

$$n \times (m \times k) = (n \times m) \times k,$$

$$(n + m) \times k = (n \times k) + (m \times k).$$

The third and fourth properties are again commutativity and associativity, but now of multiplication, and the last rule is called a *distributive* rule.

These properties can all be proved by induction, usually not directly, but by first proving all kinds of auxiliary properties by induction, after which the property itself is proven using these auxiliary properties. These are only equalities; there are also a lot of other properties to formulate which always apply. Like: if n and m are natural numbers and $n \times m = 1$, then $n = 1$

and $m = 1$. It is not the intention of this book to investigate and cover all of these characteristics; from now on we will just use such generally known valid properties when needed.

Divisors and prime numbers

Now that we have defined multiplication, we can define what a *divisor* is: we say that a natural number m is a divisor of a natural number n if a natural number k exists such that $n = k \times m$. So 2 is a divisor of 6, because $6 = 3 \times 2$. Since every n can be written as $n = n \times 1 = 1 \times n$, for every natural number n both 1 and n are divisors of n. A number n having no divisors other than 1 and n is called a *prime number*, or shortly *prime*. There are solid reasons not to call 1 a prime number, which we will come to later. The prime numbers below 100 are:

$2, 3, 5, 7, 11, 13, 17, 19, 23, 29, 31, 37, 41, 43, 47, 53, 59, 61, 67, 71, 73, 79, 83, 89, 97.$

We will show the following important properties about prime numbers:

- there are infinitely many of them, and
- every natural number can be written in exactly one way as a product of prime numbers, in order from smallest to largest.

But first, let's consider some algorithmic questions: how to determine whether a given number is prime, and how to find all primes below a given number, such as below 100 above? Determining whether a number n is prime is equivalent to looking for numbers m and k, both >1, satisfying $n = m \times k$. If such m and k can be found then n is not prime, and if such m and k can be shown not to exist, then n is prime. The easiest way is to check for all numbers $m = 2, 3, \ldots$ whether m is a divisor of n. You can do that by applying long division as you may have learned in school. Here you may skip some numbers, for instance, you don't have to try $m = 4$ if you already know that 2 is not a divisor of n, because if $n = 4 \times k$ then also $n = 2 \times (2 \times k)$. More general, you may skip any m that is not prime. But how far should you go? Note that if $n = m \times k$, then you may assume that $m \leq k$, otherwise just swap m and k. But if $n = m \times k$ and $m \leq k$, then $m \leq \sqrt{n}$, so when looking for divisors m in this way you may stop when $m > \sqrt{n}$. For instance, if you want to determine that 2857 is a prime number in this way, which is indeed the case, all you have to do is check for the prime numbers m from 2 to 53 that m is not a divisor of 2857, since for the next prime number 59 we have $59^2 = 3481 > 2857$, so $59 > \sqrt{2857}$. There are more sophisticated ways to determine whether a single number is a prime number, but we won't go into that here.

Instead we will now see how to find all prime numbers under a certain number, as above under 100. One way is to use the above method successively for all numbers $n = 2.3, \ldots$ to determine if n is prime, but it can be done much faster. This much faster way is known as *sieve of Eratosthenes*, for the first time described by the Greek scientist Eratosthenes (ca. 276–195 BC). This indeed indicates that the concept of prime number is very old. Eratosthenes is most famous for his amazingly accurate calculation of circumference of the earth by measuring the highest position of the Sun on the longest day of the year at various places in Egypt (he lived in Alexandria). Note that this assumes the spherical shape of the earth and its relationship to the Sun, which was widely doubted long afterwards.

Back to prime numbers. Assume the goal is to find all primes below some number n. Start by the sequence of all numbers from 2 to n. Then the following is done repeatingly: the first element of the sequence is stated to be a prime number, and then it is removed together with all its multiples that appear in the sequence. So in the first round, 2 is determined to be prime, and 2 and all its multiples (i.e., all even numbers) are removed from the sequence. For $n = 100$ it looks like this:

$$3, 5, 7, 9, 11, 13, 15, 17, 19, 21, 23, 25, \ldots, 97, 99.$$

In the second round, we determine that 3 is prime, and we remove 3 and all its multiples from the sequence:

$$5, 7, 11, 13, 17, 19, 23, 25, 29, 31, 35, 37, \ldots, 95, 97.$$

In the third round, we determine that 5 is prime, and we remove 5 and all its multiples from the sequence:

$$7, 11, 13, 17, 19, 23, 29, 31, 37, 41, 43, 47, 49, 53, 61, 67, 71, 73, 77, 79, 83, 89, 91, 97.$$

In the fourth round, we determine that 7 is prime, and we remove 7 and all its multiples from the sequence:

$$11, 13, 17, 19, 23, 29, 31, 37, 41, 43, 47, 53, 61, 67, 71, 73, 79, 83, 89, 97.$$

How long should this process be continued? As soon as the square of the first element in the sequence is $>n$, the process stops and the sequence consists of all remaining prime numbers under n. In this example for $n = 100$ the row starts with 11, and the square of that is >100, so we're done. And indeed: together with the numbers 2, 3, 5 and 7, which were already determined to be prime numbers, the sequence consists of exactly all prime numbers under 100 as listed above.

That this sieve of Eratosthenes works well, that is, yields exactly all prime numbers below n, will not be proved in detail here, but we give the idea. Every number that ever appears first in the sequence should be a prime number because otherwise it would already have been removed. Any other number that is removed is a multiple of a number, hence not a prime number. Removing multiples of a prime always starts with its square, because all smaller multiples have already been removed in previous rounds. And if that square is already >n, there is nothing left to remove, and the sequence consists exactly of the remaining prime numbers under n.

Now we arrive at the promised important properties of prime numbers. We are going to prove this properly, and then it is appropriate to call these properties *theorems* . Such a theorem is a formally stated property followed by its *proof*: a series of arguments together showing in detail that the property holds. For the sake of clarity, it is useful to mark the end of the proof: the last argument concluding that the property holds. In the past, this was often indicated by the letters *QED*, which is an abbreviation of *quod erat demonstrandum*, which is Latin for *what was to be demonstrated*. Nowadays it is usually indicated with a block □, just like we will do.

If a theorem becomes somewhat complicated, it may be useful for a clear proof to first prove an auxiliary theorem. Such an auxiliary theorem is called *lemma*. The usefulness of lemmas is twofold:

- If a proof is very complicated, it is easier to present and prove partial properties separately as lemmas and then use these lemmas in the final proof.
- If the same property is used in several proofs, one makes a lemma out of it. The lemma is only proved once, while it will be used several times.

For this last reason, it is also useful for our prime numbers to use a lemma.

Lemma 2.1 *If $n > 1$ is a natural number, then n has at least one divisor being > 1, and the smallest of these is prime.*

Proof: Clearly n is a divisor of n with $n > 1$, so there is at least one divisor being > 1. Now let p be the least divisor of n with $p > 1$, let's say $n = p \times q$.

Suppose that p is not prime, then $p = k \times m$ with $k, m > 1$. But then $k < p$, and because $n = k \times m \times q$ the number $k > 1$ is also a divisor of n, contradicting the assumption that p is the smallest. So p is a prime number. $\square$

Theorem 2.1 *There are infinitely many prime numbers.*

Proof: Assume there are only finitely many prime numbers, let's say n, and let's call them $p_1, p_2, p_3, \ldots, p_n$. Now we consider the number

$$N = p_1 \times p_2 \times p_3 \times \cdots \times p_n + 1.$$

Now let p be the smallest divisor of N with $p > 1$. Then p is a prime number according to Lemma 2.1. But then p is one of the numbers $p_1, p_2, p_3, \ldots, p_n$. Now let q be the product of all other prime numbers. Then $N = p \times q + 1$. But p was also a divisor of N, so $p \times k = N$ holds for a natural number k. But then we have $p \times k = N = p \times q + 1$. From these properties we obtain $p \times (k - q) = 1$, so 1 has a divisor $p > 1$, which is not possible. This contradiction indicates that the assumption that there are finitely many primes is incorrect, so there are infinitely many primes. $\square$

So indeed by this proof, using the lemma that we already proved, we have proved the theorem stating that there are infinitely many prime numbers. We promised to show one more important property of prime numbers: every natural number can be written in exactly one way as a product of primes, in order from smallest to largest. Here prime numbers may occur more than once, for example $12 = 2 \times 2 \times 3$.

Theorem 2.2 *For every natural number $n > 1$ there exists exactly one number $k > 0$ and exactly one sequence $p_1, p_2, \ldots, p_k$ of k prime numbers such that*

$$p_1 \leq p_2 \leq \cdots \leq p_k, \quad and \quad n = p_1 \times p_2 \times \cdots \times p_k.$$

Proof: Let p_1 be the least divisor > 1 of n. Then p_1 is a prime according to Lemma 2.1. If $p_1 = n$ then we have $k = 1$ and we are done. If not, then $n = p_1 \times m$ for $m > 1$. Let p_2 be the least divisor > 1 of m. If $p_2 = m$ then we have $k = 2$ and we're done. Because p_1 was the least divisor of n, then $p_1 \leq p_2$ holds.

We continue this process, and each time we are done or we find a new prime number in our sequence. Because n is finite, this process cannot continue forever, hence it stops in a situation with $p_1 \leq p_2 \leq \cdots \leq p_k$ and

$n = p_1 \times p_2 \times \cdots \times p_k$. So this shows that every $n > 1$ can be written in the desired format.

It is more difficult to show that this can only be done in exactly one way; we only give a sketch. Suppose there is a number $n > 1$ that can be written in two different ways: $n = p_1 \times a = q_1 \times b$, where a is a product of primes $\geq p_1$ is, and b is a product of prime numbers $\geq q_1$. Then choose n to be the smallest number with this property. Then $p_1 \neq isq_1$, because otherwise n/p_1 can also be written in two different ways as a product of prime numbers, and $n/p_1 < n$. Assume that $p_1 < q_1$, otherwise you swap them. Then $p_1 \times q_1 < n$, and $n - p_1 \times q_1 = p_1 \times (a - q_1) = q_1 \times (b - p_1)$. Because $n - p_1 \times q_1 < n$ the number $n - p_1 \times q_1$ can be written in exactly one way as a product of prime numbers, in which both p_1 and q_1 are included. So we can write $n - p_1 \times q_1 = p_1 \times q_1 \times c$, and so $n = p_1 \times q_1 \times (c+1)$. But then $b = p_1 \times (c+1)$, while b was a product of primes $\geq q_1$, so all $> p_1$, which contradicts the fact that $b < n$ can only be written as a product of primes in one way. $\square$

Theorem 2.2 shows that every natural number can be written in a unique way as a product of primes. If you think of multiplication as operation to make larger numbers, then in this way the prime numbers form the building blocks for making all natural numbers.

Products of several times the same number can be expressed as *exponentiation* :

$$p^m = \underbrace{p \times p \times \cdots \times p}_{m \text{ factors } p}.$$

So we have $p^1 = p$, $p^2 = p \times p$, etc. Instead of $72 = 2 \times 2 \times 2 \times 3 \times 3$ we can now write shorter $72 = 2^3 \times 3^2$. We also denote $p^0 = 1$, and will also use this exponent notation for numbers other than primes.

Using this notation and combining all equal prime numbers in Theorem 2.2, we reformulate this theorem to the following theorem:

Theorem 2.3 *For every natural number $n > 1$ there exists exactly one number $k > 0$ and k primes $p_1, p_2, \ldots, p_k$ and k natural numbers $m_1, m_2, \ldots, m_k > 0$ such that*

$$p_1 < p_2 < \cdots < p_k, \quad \text{and} \quad n = p_1^{m_1} \times p_2^{m_2} \times \cdots \times p_k^{m_k}.$$

Note that the number k in Theorem 2.3 is different from the number k in Theorem 2.2, for instance, for $n = 72$ we have $k = 5$ in Theorem 2.2 and $k = 2$ in Theorem 2.3.

When we defined prime numbers it was not yet clear whether 1 should be a prime number or not. Since 1 has no divisors other than 1 and itself, you might tend to call 1 prime. But if you do so, Theorem 2.2 does not hold any more, since then there would be a lot of ordered prime number products with the same result, for example $6 = 2 \times 3 = 1 \times 2 \times 3 = 1 \times 1 \times 2 \times 3 = \cdots$

This could be fixed by adding the requirement in Theorem 2.2 that $p_1 > 1$, but then not the primes but only the primes > 1 form the unique building blocks for all natural numbers, and 1 inevitably plays a very different role than the other prime numbers. That is why all mathematicians who work with prime numbers nowadays agree that 1 is not considered to be a prime number, and we follow this agreement.

The notation of the natural number n in Theorem 2.3 is called the *prime factorization* of n.

About prime numbers much more can be said. The theorem that there are infinitely many of them can be refined very much by deriving precise formulas for the number of prime numbers below n, as a function of n. And a lot is still unknown about prime numbers. A striking example of this is *Goldbach's conjecture*: every even number > 2 can be written as the sum of two primes. Indeed:

$$4 = 2 + 2, \ 6 = 3 + 3, \ 8 = 3 + 5, \ 10 = 3 + 7, \ 12 = 5 + 7, \ 14 = 3 + 11,$$

$$16 = 3 + 13, \ 18 = 5 + 13, \ 20 = 7 + 13, \ 22 = 3 + 19, \ldots$$

Often it can be done in many ways, such as $22 = 3 + 19 = 5 + 17 = 11 + 11$, but no even number > 2 has ever been found where it cannot be done. In fact, it has been proved that all even numbers from 4 to 4×10^{18} can been written as the sum of two primes. But no one has ever been able to prove that it also applies to all larger even numbers, despite immense efforts and many partial results.

This chapter is approaching its end, so let's summarize what we have seen. We started by a very simple definition of natural numbers on which operations like $+$ and $\times$ were inductively defined. The we defined prime numbers and investigated some basic properties. The goal of this was not only conducting this particular investigation but also showing the flavor of setting up some mathematics, by giving proper definitions, formulating and proving theorems, and structuring this process by presenting auxiliary properties as lemmas. Such a lemma may be used for several times, in our case, in the proofs of Theorems 2.1 and 2.2.

Before closing this chapter with the promised challenge, let's return to the paint pot problem: the challenge of Chapter 1. A good strategy is that if you're stuck on a challenge, then simplify the problem a bit and see if you can say something about it. If you're trying to do that with the paint pot problem, an obvious step is to reduce the number of colors from four to three. In that case we only have the letters a, b, c and d where a represents an empty paint pot, and b, c and d represent the three colors of paint. Then the rules are

$$aba = bab, \; aca = cac, \; ada = dad, \; bc = cb, \; bd = db, \; cd = dc,$$

and the question is whether it is possible to start with a sequence of letters starting in bcd, and just apply these rules to end up with a sequence starting in a. It turns out that it is indeed possible:

$$
\begin{aligned}
bc\underline{dad}cabacda &= bcad\underline{aca}bacda &= bcadca\underline{cb}acda &= bcadcab\underline{cac}da &= \\
bcadcabac\underline{ada} &= bcadc\underline{ab}acdad &= bca\underline{dcb}abcdad &= b\underline{cac}dbabcdad &= \\
bac\underline{ad}babcdad &= bacabda\underline{bcd}ad &= bacabda\underline{bd}cad &= bacab\underline{dad}bcad &= \\
bac\underline{aba}dabcad &= bac\underline{ba}babcdad &= \underline{bab}cabdabcad &= abacabdabcad.
\end{aligned}
$$

Here, the part of the sequence that is modified by a rule is always underlined. This is a special solution: it starts with the shortest possible sequence having a solution, and this solution has the smallest possible number of steps. First this solution was found by hand, just after puzzling for a while. Variants on this are also possible, for instance by doing some swappings some steps earlier or later. A similar solution has also been found by a computer program, and from the search process of the program we know that this is the smallest and shortest solution.

This doesn't solve the real paint pot problem with four colors, but this solution for three colors already indicates that if there is a solution for four colors, it will, in any case, have to be longer than the above solution for three colors. To be continued!

Challenge: number of divisors

This chapter was about natural numbers, and special natural numbers such as prime numbers and divisors of numbers. This is exactly what this challenge is about.

Challenge:

Consider numbers n that have exactly 109 divisors apart from 1 and n. Show that there are infinitely many of them, and that all of them are >600 billion.

More complicated numbers

In Chapter 2, we considered natural numbers. But these do not include more complicated numbers like -3, a negative number, or $\frac{3}{4}$, a fraction, not to mention $\sqrt{2}$ or π. If your world is limited to natural numbers, you can just say that these more complicated numbers don't exist, and you don't have to worry about them.

But there are also situations where it is useful to be able to find a number x with $4 + x = 1$, or a number y with $4 \times y = 3$, or even a number z with $z \times z = 2$. Within the natural numbers such numbers x, y, z do not exist, but in richer number systems you do have numbers x, y, z with these desired properties, namely $x = -3$, $y = \frac{3}{4}$ and $z = \sqrt{2}$. In this chapter, we will see how we can build these richer number systems using only our natural numbers as building blocks. For the different sets of numbers that we are going to look at, it is useful to have a fixed notation. The usual notation for the set of natural numbers is $\mathbb{N}$, which we will use here too. This is a capital N with a double bar, the N being the N of natural numbers.

Integer numbers

The integer numbers, or shortly integers, consist of the natural numbers extended by negative numbers. The notation for the set of integers is $\mathbb{Z}$, derived from the German word *Zahlen*. So we want to obtain $\mathbb{Z}$ by adding

$$-1, -2, -3, -4, \ldots$$

to $\mathbb{N}$, so there are infinitely many numbers to add. And we will not add them one by one, but give a construction to do that at once.

DOI: 10.1201/9781003466000-3

We did not formally define the $-$ subtraction operation for natural numbers. One reason for that was the annoying complication that you don't always get a natural number when you put in two natural numbers. But now we are going to use this complication as a feature instead: we will define $n - m$ for all natural numbers n, m, and will define the integers as being the set of results. Let's show the intended numbers $n - m$ in a picture:

$$
\begin{array}{ccccc}
\vdots & \vdots & \vdots & \vdots & \\
0-3 & 1-3 & 2-3 & 3-3 & \cdots \\
0-2 & 1-2 & 2-2 & 3-2 & \cdots \\
0-1 & 1-1 & 2-1 & 3-1 & \cdots \\
0-0 & 1-0 & 2-0 & 3-0 & \cdots
\end{array}
$$

Imagine this picture with the dots at the top continued infinitely upwards, and with the dots on the right continued infinitely to the right. Then we have two main observations, consisting of good news and bad news.

The good news is that all integers are here: any integer can be written as $n - m$ for natural numbers n, m.

The bad news is that each of those integers appears multiple times, even infinitely often, for instance

$$2 = 2 - 0 = 3 - 1 = 4 - 2 = 5 - 3 = \cdots$$

Because of this bad news, we're not done yet. What we want to do is that we want to group together all the expressions $n - m$ in this picture that yield the same number. If we succeed, we can define the integers as the resulting groups.

When do $n - m$ and $n' - m'$ end up in the same group? If they result in the same integer value. But we don't have any integers yet, because we're just defining them, and we haven't defined the $-$ operator either. But instead of $n - m = n' - m'$, we can equivalently write $n + m' = n' + m$, only using natural numbers and addition that were defined properly.

So what we do is this: we consider all the expressions $n - m$ as they appear in the infinitely continued picture above, for all natural numbers n, m, and we put $n - m$ and $n' - m'$ together in a group whenever $n + m' = n' + m$. Now we define the integers as the resulting groups.

What does such a group look like? It turns out to consist of all expressions $n - m$ that lie on a diagonal line as shown in the following picture:

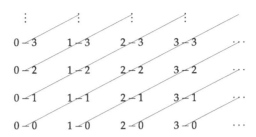

Here the number 0 corresponds to the group consisting of 0−0, 1−1, 2−2,..., being the middle diagonal line in the picture. The number 1 corresponds to the group consisting of 1 − 0, 2 − 1, 3 − 2,..., being the diagonal line in the picture to the right of the middle one. The number 2 corresponds to the group consisting of 2 − 0, 3 − 1, 4 − 2,.... That is the diagonal line in the picture to the right of the two diagonals we already considered. This continues, by which every natural number corresponds to such a diagonal line.

But there are even more groups or diagonal lines. To the left of the one for the number 0, we have the group consisting of 0−1, 1−2, 2−3,..., which is the number that we denote by −1. To the left of that again we have the group consisting of 0 − 2, 1 − 3, 2 − 4,..., which is the number that we denote by −2. And so on, by which all integers are defined and thus the set $\mathbb{Z}$.

This may look as a quite complicated way of describing integers. But this way of defining new concepts is applied very often in mathematics and computer science. The general shape is as follows: first define a set that is too large a set. Then this too large set is split into groups of elements that should be equivalent and then defining the new objects as being these groups. We will use the same idea later to define the rational numbers and the real numbers. In mathematics, such division into groups is called a *partition*, the notion of being in the same group is called an *equivalence relation*.

Let's show the construction we just gave in a little more mathematical notation. If we have two sets A and B, then we denote the set of pairs (a, b) for which a is in A and b is in B by $A \times B$. This is called the *Cartesian product* of A and B, named after René Descartes (1596–1650), a French philosopher and mathematician, who studied and worked in the Netherlands a great part of his life. We started with the expressions $n − m$ where n and m were natural numbers, and the '−' had no meaning yet. So this was actually $\mathbb{N} \times \mathbb{N}$. And on this we have created the equivalence relation $\sim$ for which $(n, m) \sim (n', m')$ holds exactly if $n + m' = n' + m$. And then we defined the set $\mathbb{Z}$ to consist of the resulting groups. This is denoted as follows:

$$\mathbb{Z} = (\mathbb{N} \times \mathbb{N})/\sim.$$

Using the division symbol '/' in this way is also called: taking the *quotient* of the equivalence relation $\sim$.

We may think of $\mathbb{Z}$ as an extension of $\mathbb{N}$ by identifying each n in $\mathbb{N}$ with $n - 0$ in $\mathbb{Z}$: in this way every natural number, being an element of $\mathbb{N}$, is an element of $\mathbb{Z}$ too, so is also an integer number.

The operations $+$ and $\times$, i.e. addition and multiplication, can also be defined on integers. On integers now *subtraction* can be introduced in a way that, in contrast to the natural numbers, it is always defined. More precisely: for all integers n, m exactly one integer $n - m$ exists for which $m + (n - m) = n$.

We write $-n$ as a shorthand notation for $0 - n$. A property that can be derived for this is $(-n) \times (-n) = n \times n$. Further details are omitted here.

Rational numbers

The operation of subtraction is a kind of *inverse* of addition, because of $m + (n - m) = n$. We may wonder if we can do something similar with multiplication. We still remember the result from primary school: we speak of *division*, and we write the result as n/m, or $\frac{n}{m}$. Just like we have $m + (n - m) = n$ for addition and subtraction, now we expect the property $m \times (n/m) = n$ to hold. For every n,m, is it possible to find such n/m with this property? If m is a divisor of n, yes. And if $m = 0$ then it will never work if $n \neq$ is0, because then m multiplied by something is always 0, not equal to n. But if m is not equal to 0, it may be useful to allow to divide n by m. For instance, if we want to divide 3 by 4 we can say we can't (and in the natural and integer numbers indeed we can't), but we can also say that we get $\frac{3}{4}$: a new number in our new number system with a clear practical use: a rope of length 3 indeed can be divided into 4 equal pieces each having length $\frac{3}{4}$. In this way, we obtain the *rational numbers*. The set of rational numbers is denoted by $\mathbb{Q}$, derived from the 'Q' of 'quotient'. Now that we already have the sets of natural numbers $\mathbb{N}$ and integers $\mathbb{Z}$, we now use them to formally define the set of rational numbers $\mathbb{Q}$, in a similar way to how we defined $\mathbb{Z}$ from $\mathbb{N}$.

A rational number is of the form $\frac{n}{m}$, and we call this representation a *fraction*, where n, the *numerator*, is an integer, and where m, the *denominator*, cannot be 0. When allowing both positive and negative values for the numerator, it is not necessary to also allow negative denominators. Let's denote the set of natural numbers > 0 by $\mathbb{N}^+$. We now see a fraction $\frac{n}{m}$ as a pair of (n, m) of numbers, with (n, m) in $\mathbb{Z} \times \mathbb{N}^+$, so the enumerator n may be any integer, and the denominator m is some natural number > 0. As with the integer numbers, the intended fractions can be shown in a picture:

$$
\begin{array}{ccccccc}
\vdots & \vdots & \vdots & \vdots & \vdots & \vdots & \vdots \\
\cdots \quad \frac{-3}{4} & \frac{-2}{4} & \frac{-1}{4} & \frac{0}{4} & \frac{1}{4} & \frac{2}{4} & \frac{3}{4} \quad \cdots \\
\cdots \quad \frac{-3}{3} & \frac{-2}{3} & \frac{-1}{3} & \frac{0}{3} & \frac{1}{3} & \frac{2}{3} & \frac{3}{3} \quad \cdots \\
\cdots \quad \frac{-3}{2} & \frac{-2}{2} & \frac{-1}{2} & \frac{0}{2} & \frac{1}{2} & \frac{2}{2} & \frac{3}{2} \quad \cdots \\
\cdots \quad \frac{-3}{1} & \frac{-2}{1} & \frac{-1}{1} & \frac{0}{1} & \frac{1}{1} & \frac{2}{1} & \frac{3}{1} \quad \cdots
\end{array}
$$

In this picture, there are several pairs (n, m) that denote the same value $\frac{n}{m}$ of the corresponding rational number, for instance $\frac{1}{2} = \frac{2}{4} = \frac{3}{6} = \cdots$. More general, we have $\frac{n}{m} = \frac{n'}{m'}$ exactly if $n \times m' = n' \times m$. So similar as we did for defining integers, on the pairs (n, m) we now define the new equivalence relation $\sim$ with $(n, m) \sim (n', m')$ exactly if $n \times m' = n' \times m$. By taking the quotient, we now get the definition of $\mathbb{Q}$, being the set of rational numbers:

$$
\mathbb{Q} = (\mathbb{Z} \times \mathbb{N}^+)/\sim .
$$

This is indicated in the following picture in which the points that are in the same group are connected by a line:

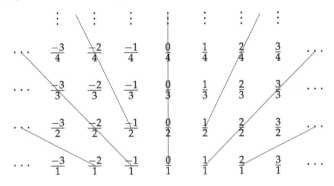

This picture only shows a finite initial part, and not all lines are drawn. In the full picture, for every $\frac{n}{m}$ where n and m have no divisors in common, there would be a line through $\frac{k \times n}{k \times m}$ for every $k > 1$, because they all give the same value $\frac{n}{m}$, and they do indeed line up nicely in the picture.

We can think of this set $\mathbb{Q}$ as an extension of $\mathbb{Z}$ by identifying each n from $\mathbb{Z}$ with $\frac{n}{1}$. The operators $+$, $-$ and $\times$ from $\mathbb{Z}$ can be extended to operators on $\mathbb{Q}$. We will not elaborate this in detail, but these operations behave exactly like you learned in school. For instance, in order to add fractions you first have to make the denominators equal, and so on. Many properties, like $+$ and $\times$ being commutative and associative, that hold in $\mathbb{N}$ and $\mathbb{Z}$, also hold in this extended setting. But also some new properties show up. For instance, for every q, r in $\mathbb{Z}$ with r unequal to 0, exactly one s in $\mathbb{Q}$ exists such that $r \times s = q$. We also write s as q/r or $\frac{q}{r}$: in this way we have a new operation '/', *divide*. This is very similar to the new operation '$-$', subtraction, that we obtained when extending $\mathbb{N}$ to $\mathbb{Z}$. had the new operation '$-$', subtraction. Also all the previous rules we had, such as the commutative, associative and distributive rules, still apply in $\mathbb{Q}$.

On $\mathbb{N}$, $\mathbb{Z}$ and $\mathbb{Q}$ the relations $<$, $\leq$, $\geq$ and $>$ can be defined. In the previous chapter we already used these relations on $\mathbb{N}$ when considering prime numbers. These can be defined formally by induction as a relation on $\mathbb{N}$. Then these can be extended to $\mathbb{Z}$ and $\mathbb{Q}$. Because these relationships are so well known, and everything you expect also applies, we will not elaborate it here. But it is interesting to note that $\mathbb{Q}$ has a new property that does not hold in $\mathbb{N}$ and $\mathbb{Z}$, namely that for every q, r in $\mathbb{Q}$ with $q < r$ there is an s in $\mathbb{Q}$ that is in between, i.e.

$$q < s \text{ and } s < r.$$

In $\mathbb{N}$ and $\mathbb{Z}$ this does not hold. For instance, $1 < 2$ holds, but there is no natural or integer number in between 1 and 2. In $\mathbb{Q}$ such a number exists: $\frac{3}{2}$. More generally, for every q, r in $\mathbb{Q}$ with $q < r$ it can be shown that the number $s = \frac{q+r}{2}$ is in between, that is, $q < s$ and $s < r$ satisfy.

Later in this book we will see how rational numbers play an important role in the behavior of turtle figures.

Real numbers

You can do a lot with the rational numbers. We just saw, for example, that there is another rational number between every pair of distinct rational numbers. When measuring distances, this is a very useful, since this shows that distances may be refined indefinitely to ever smaller distances. But is it possible to measure the length of the diagonal of a unit size square, as a rational number? More precisely, does a rational number q exist with $q \times q = 2$, that is, a number that represents $\sqrt{2}$? The following statement provides the answer.

Theorem 3.1 *There is no q in $\mathbb{Q}$ with $q \times q = 2$.*

Proof: Suppose there is such a q. Write $q = \frac{n}{m}$ with n in $\mathbb{Z}$ and m in $\mathbb{N}^+$. Then from $q \times q = 2$ it follows that $\frac{n}{m} \times \frac{n}{m} = 2$, hence $n \times n = m \times m \times 2$. If n is negative, we replace n by $-n$, because due to $(-n) \times (-n) = n \times n$ the property $n \times n = m \times m \times 2$ still holds. It also follows from $n \times n = m \times m \times 2$ that n is not 0 or 1, so n is a natural number with $n > 1$. According to Theorem 2.3 from the previous chapter it now follows that there is a number $k > 0$ and k primes $p_1, p_2, \ldots, p_k$ and k natural numbers $m_1, m_2, \ldots, m_k > 0$ such that

$$p_1 < p_2 < \cdots < p_k$$

and

$$n = p_1^{m_1} \times p_2^{m_2} \times \cdots \times p_k^{m_k}.$$

It follows

$$n \times n = p_1^{2m_1} \times p_2^{2m_2} \times \cdots \times p_k^{2m_k}.$$

Because $n \times n = m \times m \times 2$, we conclude that n is even and $p_1 = 2$. The prime number 2 thus appears in the prime factor decomposition of $n \times n$ with an *even* power $2m_1$. If we apply Theorem 2.3 to m in the same way, we get that the prime number 2 occurs in the prime factorization of $m \times m \times 2$ with an *odd* power. This contradicts the fact that according to Theorem 2.3 the prime factorization of $n \times n = m \times m \times 2$ is unique. $\square$

So within $\mathbb{Q}$ the number $\sqrt{2}$ does not exist. Do we think that $\sqrt{2}$ does exist? In the world of measuring distances is $\sqrt{2}$ a value that you encounter, namely as the length of the diagonal of a square with side 1. In ancient Greece, Hippasus (ca. 530–450 BC), a student of Pythagoras (ca. 570–495 BC), was the first to observe Theorem 3.1. It seems that this heavily shocked some followers of Pythagoras. There are several stories about the reactions of these followers, some of which causing that Hippasus did not survive.

Just like we extended the natural numbers to integers in order to include negative numbers, and we extended the integers to rational numbers in order to include fractions, we now want to extend the rational numbers to something that does contain numbers like $\sqrt{2}$. We could do this in small steps, by first adding $\sqrt{2}$, then $\sqrt{3}$, being the *zeros* of the *polynomials*

$$x^2 - 2 \quad \text{and} \quad x^2 - 3$$

and then zeros of more general polynomials

$$x^n + a_{n-1}x^{n-1} + a_{n-2}x^{n-2} + \cdots + a_1x + a_0$$

where $a_0, a_1, \ldots, a_{n-1}$ are rational numbers. This way we get the *algebraic numbers*. There is a lot to tell about this, in fact my whole PhD thesis is about the structures you get in the first steps doing so. But then we are not yet finished: it turns out that the number $\pi = 3.141592\ldots$, being the circumference of a circle with diameter 1, still cannot be obtained. Therefore, here we are not going to define polynomials and algebraic numbers, but we choose to to make a bigger step in one go by which a number like π can also be obtained. The extension of the rational numbers that we get in this way are the *real numbers*.

In doing so, we will use the important concept of a *limit*. We say that an infinite sequence of numbers

$$a_0, a_1, a_2, a_3, a_4, \ldots$$

has limit L if for large values of n the value of a_n is very close to L. Here 'very close to' means: as close as you want. This is defined formally as follows. It starts by taking a very small number $\epsilon > 0$, for example 0.00001 or 0.00000001. We could also call this a or b or whatever other letter, but in mathematics it is common to use the Greek letter ϵ for such a very small number, so that's what we do here. Then we want that a_n is very close to L, that is

$$L - \epsilon < a_n < L + \epsilon,$$

if n is very large. What do we mean by 'very large'? That means there is some N such that this holds for all $n > N$. In summary, we now have the definition of limit:

A sequence $a_0, a_1, a_2, a_3, a_4, \ldots$ of numbers has *limit L* if for every $\epsilon > 0$ some number N exists such that for all $n > N$ the following holds:

$$L - \epsilon < a_n < L + \epsilon.$$

We did not specify what kind of numbers $a_0, a_1, a_2, a_3, a_4, \ldots$ are. For the time being, let's assume that they are rational numbers. Now we come to a surprising property: we have seen that $\sqrt{2}$ itself is not a rational number, but $\sqrt{2}$ can be obtained as the limit of a sequence of rational numbers. In our usual decimal notation, we have

$$\sqrt{2} = 1.41421356237\cdots$$

One way to choose a suitable sequence is by defining a_0 as this number with 0 decimal places, a_1 with one decimal place, a_2 with two decimal places, and so on, so

$$a_0 = 1, \ a_1 = 1.4, \ a_2 = 1.41, \ a_3 = 1.414, \ a_4 = 1.4142, \ldots$$

All these numbers are rational:

$$a_0 = \frac{1}{1}, \ a_1 = \frac{14}{10}, \ a_2 = \frac{141}{100}, \ a_3 = \frac{1414}{1000},$$

etcetera. But its limit will be $\sqrt{2}$: for every $\epsilon > 0$ we can find a N with $\epsilon > \frac{1}{10^N}$. For every $n > N$ the first N decimal places of a_n and $\sqrt{2}$ are equal to each other, so

$$L - \epsilon < \sqrt{2} - \frac{1}{10^N} < a_n < \sqrt{2} + \frac{1}{10^N} < L + \epsilon.$$

This shows that $\sqrt{2}$ is indeed the limit of this particular sequence of rational numbers.

At this moment one serious problem remains: we did not yet define $\sqrt{2}$ as a number, as it is outside the rational numbers, being the richest class of numbers we considered until now. We want to establish having a limit without referring to that limit itself. This is not possible with the definition of limit itself because the limit L is used there. The definition of limit indicates that for n large enough, the value of a_n is very close to L. But instead we now will require that for all n, m that are both very large, the values of a_n and a_m are very close to each other. Because the idea of this goes back to the Frenchman Augustin Louis Cauchy (1789–1857), the resulting sequences are called *Cauchy sequences*:

> A sequence $a_0, a_1, a_2, a_3, a_4, \ldots$ of numbers is called a *Cauchy sequence* if for every $\epsilon > 0$ there exists a number N such that for every $n, m > N$ the following holds:
>
> $$-\epsilon < a_n - a_m < \epsilon.$$

Now we will define the real numbers based on Cauchy sequences. Roughly speaking, the real numbers will consist of the limits of Cauchy sequences. As a first attempt, the real numbers will be the Cauchy sequences themselves. But here we have to take a quotient again, because there are a lot of Cauchy sequences having the same limit, so representing the same number, for example $\sqrt{2}$. We may take the above sequence for $\sqrt{2}$, but if we replace a_0

with 17 and a_1 with 83 and leave the rest of the a_n untouched, we have another Cauchy sequence also representing $\sqrt{2}$. Just like we did when defining integers and rational numbers, we now define an equivalence relation for taking the quotient:

For two Cauchy sequences $a = a_0, a_1, a_2, a_3, a_4, \ldots$ and $b = b_0, b_1, b_2, b_3, b_4, \ldots$ then $a \sim b$ holds exactly if

$$a_0, b_0, a_1, b_1, a_2, b_2, a_3, b_3, a_4, \ldots$$

is also a Cauchy sequence.

The idea is that if a and b are two Cauchy sequences with the same limit, then the new sequence has the same limit and is also a Cauchy sequence, but if a and b are two Cauchy sequences with different limits L_a and L_b, then the new row will keep swabbing around those L_a and L_b, and has no limit itself and is not a Cauchy sequence.

Writing C for the set of all Cauchy sequences of rational numbers, we now define the set $\mathbb{R}$ of *real numbers* as

$$\mathbb{R} = C/\sim .$$

Without giving proofs, we give some important properties of $\mathbb{R}$:

- On $\mathbb{R}$ we have the operations $+$, $-$, $\times$ and $/$, just like we had in $\mathbb{Q}$,
- In $\mathbb{R}$ every Cauchy sequence has a limit.

So in particular we know that $\sqrt{2}$ is in $\mathbb{R}$ because that is the limit of the sequence

$$a_0 = 1, \ a_1 = 1.4, \ a_2 = 1.41, \ a_3 = 1.414, \ a_4 = 1.4142, \ldots$$

which we already discussed above. Similarly, numbers like $\sqrt{3}$ and π are also in $\mathbb{R}$.

This is not the only possible way to define $\mathbb{R}$. Another approach is based on *Dedekind cuts*, named after the German mathematician Richard Dedekind (1831–1916). A typical Dedekind cut is the set of rational numbers less than $\sqrt{2}$, but here $\sqrt{2}$ may be replaced by any other real number. A Dedekind cut is by definition a set D of rational numbers with the following properties:

- For every d in D, every rational number q with $q < d$ is also in D.
- For every d in D there is a d' in D with $d' > d$.
- There is a d in D, but there is also a rational number that is not in D.

The surprising thing is that the Dedekind cuts are precisely the sets consisting of all rational numbers $< r$ for a certain real number r. They thus fix exactly that number r, and can be used as a definition of the real numbers, without defining an equivalence relation and taking the corresponding quotient.

Complex numbers

Do we cover all kinds of useful numbers by now? Not quite yet. We extended the rational numbers $\mathbb{Q}$ to the real numbers $\mathbb{R}$ in such a way that it includes $\sqrt{2}$, the value x that satisfies $x^2 = 2$. As usual, here x^2 is a shorthand notation for $x \times x$. But is there a value x that satisfies $x^2 = -1$? If your world is limited to $\mathbb{R}$, the answer is clearly 'no': you can show that in $\mathbb{R}$ the square of each number is ≥ 0, and therefore not equal to -1. But that argument is similar to stating that $\sqrt{2}$ doesn't exist because there isn't a rational number q with $q^2 = 2$. This did not prevent us from extending the rational numbers $\mathbb{Q}$ to the real numbers $\mathbb{R}$ in which such a value does exist.

What we are going to do now is to extend $\mathbb{R}$ by a special number i where $i^2 = -1$. If you think of the real numbers, then such a number i does not exist at all, so that's why this number is called an *imaginary* number, and the notation i is chosen. The *complex numbers* are now defined as all numbers of the shape $a + bi$ where a and b are real numbers. The set of all complex numbers is denoted by $\mathbb{C}$. These complex numbers can be added, subtracted and multiplied:

$$(a + bi) + (c + di) = (a + c) + (b + d)i,$$

$$(a + bi) - (c + di) = (a - c) + (b - d)i,$$

$$(a + bi) \times (c + di) = (ac - bd) + (ad + bc)i.$$

That last expression may look a bit complicated, but is obtained by just adding $a \times c$, $a \times di$, $b \times ci$ and $bi \times di$, remembering that $bi \times di = bd \times i^2 = -bd$, using $i^2 = -1$.

Just like in $\mathbb{Q}$ and $\mathbb{R}$, in $\mathbb{C}$ also dividing by any number other than 0 can be done.

Even if you argue that such a weird number i doesn't exist, defining complex numbers and these operations can be done in this way. In many applications it is much more convenient to use complex numbers than to do without. For instance, later in this book we will see how complex numbers are useful to determine positions in fractal turtle figures.

It is not the intention of this book to deal with complex numbers extensively, but it is important to see how it is the next logical step after considering real numbers.

Are the complex numbers the end of these series of stepwise extension of sets of numbers? No: further extensions are possible, such as the *quaternions*. These are numbers of the shape $a + bi + cj + dk$ where a, b, c, d are all real numbers, and $i^2 = j^2 = k^2 = -1$. These auxiliary numbers i, j, k moreover satisfy $i \times j = k$ and $j \times i = -k$. Now multiplication $\times$ is no longer *commutative* any more: there are numbers x, y (like $x = i$ and $y = j$) where $x \times y$ and $y \times x$ are not equal to each other. Until now, in all extensions all earlier basic properties like this commutativity were preserved, but this shows that for the extension to quaternions this is not the case any more. So this is a good moment to finish our story of extending number systems.

Challenge: ten questions

Challenge:

This challenge consists of ten questions each having a natural number as its answer. The questions themselves have largely been lost, but the following is known:

- The answer to question 1 is the number of primes that are less than 20.
- The answer to question 2 is the sum of all other answers.
- The answer to question 3 is obtained by subtracting the answer to question 7 from the average of the answers to questions 8 and 10.
- The answer to question 4 is exactly 10% of the sum of the answers to questions 6 and 7.
- The answer to question 5 is the number of different rational numbers $\frac{n}{m}$ where n is an integer with $-5 < n < 5$ and m is a natural number with $m < 10$.
- The answer to question 6 is one third of the sum of the answers to questions 4 and 5.
- The answer to question 7 is half the product of the answers to questions 1 and 4.
- The answer to question 8 is the number of rational numbers x for which $n = x \times x$ satisfies $-10 < n < 10$.
- The answer to question 9 is the number of real numbers x for which $n = x \times x$ satisfies $-10 < n < 10$.
- The answer to question 10 is the number of complex numbers x for which $n = x \times x$ satisfies $-10 < n < 10$.

What is the sum of the answers to all ten questions?

Flavors of infinity

In this chapter, we will argue that there are different kinds of infinity, and that one kind is really bigger than the other, while they are both infinite. To be more precise, first it should be defined what is meant by *equal size* and *greater than*. Before doing so, we start by telling a story about a hotel.

The Hilbert Hotel

Assume that all rooms in a hotel are occupied. A new guest arrives and asks at the desk whether a room is available. Then it seems to be self-evident that no other answer is possible than 'no'. The hotel employee will have been trained to formulate this more diplomatically, along the lines of 'unfortunately I have to disappoint you', but that does not change the situation: the new guest will have to leave without having a room in this hotel.

But assume for now that it is not a finite hotel, as usual, but an infinitely large hotel in which all rooms are numbered with natural numbers such that for every natural number there is a room with that number. And now suppose that all rooms of the hotel are occupied as before, and a new guest arrives asking whether a room is available.

Now surprisingly the desk clerk has another option: he or she may ask all guests in the hotel to move up a room, so for every n the person in room n is asked to move to room $n + 1$. If they all do that at the same time, this will cause a lot of guests and stuff in the hallway for some time, but the room to which each guest has to go is free, so this is possible. And this action frees up room 0, where the fresh guest may move in now.

In this way, there is always a room for every single fresh guest. However, next not a single new guest arrives, but a bus full of new guests, and the tour leader of the bus asks at the desk whether there is still room in the hotel for all passengers of the bus. Now this is not a usual bus, but an infinitely large bus, in which all occupants are numbered by a natural number. The desk clerk is puzzled now, but then comes up with the following solution: all guests of the hotel are asked to move to the room whose number is twice as high, and that's what they do. So the guest in room 0 may stay, but all other guests have to move to another room: $1 \rightarrow 2$, $2 \rightarrow 4$, $3 \rightarrow 6$, and so on. In this way, all original guests in the hotel keep having a room, but all rooms with an odd number become vacant, and all the infinitely many occupants of the bus can enter exactly there: 0 to room 1, 1 to room 3, and so on.

This curious infinitely large hypothetical hotel that always has room for new guests is called the *Hilbert hotel*, named after David Hilbert (1862–1943). He is not only known for this story: he was one of the greatest mathematicians ever. Although in his days mathematics was already that extensive that it was hardly possible for one person to grasp all, Hilbert came a long way. He used to focus on some area of mathematics until he became a leading expert and solved several open problems there. After reaching this level, typically already after a few years, he switched to a completely different area of mathematics to start again. This he continued for the rest of his life, by which currently in a wide range of areas of mathematics many main results go back to Hilbert.

The story of the Hilbert hotel indicates a crucial difference between finite and infinite sets. To consider that in more detail first we need some new concepts.

Smaller than?

When do we call a set A smaller than a set B? For finite sets an obvious approach is to count the number of elements, and then A is smaller than B if its number of elements is smaller. We also want to talk about *less than* for infinite sets. One approach would be: remove elements from A one by one, and every time you remove an element from A, you also remove one from B. If at some moment still some elements remain in A while all elements of B have been removed, then you conclude that originally A was greater than B. For every a taken away from A in this approach, denote the corresponding removed element from B by $f(a)$. Then, if at some point A is completely emptied, for every a in A there is a corresponding element $f(a)$ in B.

Such an f that defines an associated element $f(a)$ in B for each element of A is called a *mapping* or *function* from A to B. We write this as $f : A \rightarrow B$. In general, for a function $f : A \rightarrow B$ it is allowed to have $f(a) = f(a')$ if a and a' are two distinct elements of A. But this is not allowed in our counting construction: if we remove a from A, and thereby $f(a)$ from B, and a little later a' from A and $f(a')$ from B, then $f(a')$ cannot be equal to $f(a)$, because it was already removed from B. A function $f : A \rightarrow B$ with the additional requirement that for every a and a' from A with $a \neq a'$ one has $f(a) \neq f(a')$, is called *injective*. If we have an injective map $f : A \rightarrow B$, then A is (strictly) less than B if there is at least one element of B left at the end. If there are no elements of B left, then A and B have the same size. If we don't care about whether or not there are elements left in B, we say that A *is less than or equal to B* if there is a *injective mapping* exists from A to B, denoted by

$$A \preceq B.$$

This is very similar to the notation $\leq$ for *less or equal* for numbers, but slightly different. For finite sets it coincides with $\leq$ for counting the number of elements; now we will also use this definition for infinite sets.

An important example of $A \preceq B$ is if A is a *subset* of B. That means that every element of A is also an element of B. We denote this by $A \subseteq B$. The similarity in the notations $\leq$, $\preceq$ and $\subseteq$ is on purpose. The observation now is that

Theorem 4.1 *If $A \subseteq B$ then $A \preceq B$.*

Proof: We have to show that $A \preceq B$ of two sets A, B with $A \subseteq B$, that is, there is an injective map $f : A \rightarrow B$. This is defined by $f(a) = a$ for all a in A. It is injective: if $a \neq a'$ then $f(a) \neq f(a')$ holds because $f(a) = a$ and $f(a') = a'$. $\square$

As this theorem and its proof is very obvious, it is often used without explicit reference.

The infinite sets of numbers we have seen satisfy

$$\mathbb{N} \subseteq \mathbb{Z} \subseteq \mathbb{Q} \subseteq \mathbb{R} \subseteq \mathbb{C},$$

and so according to this theorem also satisfy

$$\mathbb{N} \preceq \mathbb{Z} \preceq \mathbb{Q} \preceq \mathbb{R} \preceq \mathbb{C}.$$

Using these notations we now express a crucial difference between finite and infinite sets. For finite sets A and B, the following property holds:

if $A \subseteq B$ and $B \preceq A$, then A and B are equal to each other.

This can be seen as follows. Writing $|A|$ for the number of elements of A, and similar for B, from $B \preceq A$ it follows that $|B| \leq |A|$. So A is a subset of B having at least as many elements as B. This is only possible if A is the full set B.

Surprisingly, for infinite sets this property does not hold. We have already seen the set $\mathbb{N}^+$ of natural numbers > 0, and $\mathbb{N}^+ \subseteq \mathbb{N}$ holds. However, the map $f : \mathbb{N} \to \mathbb{N}^+ :$ defined by $f(n) = n + 1$ for every n in $\mathbb{N}$ is injective, since from $n + 1 = n' + 1$ we may conclude $n = n'$. So $\mathbb{N}^+ \subseteq \mathbb{N}$ and $\mathbb{N} \preceq \mathbb{N}^+$, while $\mathbb{N}$ and $\mathbb{N}^+$ are really two distinct sets. This shows that the above property does not hold. In fact this is exactly what was exploited in the Hilbert story: by shifting each element one place by this f, the first element of $\mathbb{N}$ becomes free, thus making the room available for the new incoming hotel guest.

Equal size?

Two sets A and B are called *equinumerous* if there is both an injective map from A to B and an injective map from B to A. This is denoted by

$$A \simeq B.$$

This is by definition exactly the case if both $A \preceq B$ and $B \preceq A$ hold. We may think of equinumerous as having the same size.

That may seem a bit abstract, but if you want to explain why two sets with three elements each, let's say A consisting of the numbers $0, 1, 2$ and B consisting of the names Huey, Dewey and Louie have the same size, this can be done by using an injective map $f : A \to B$ and an injective map $g : B \to A$. For instance, choose $f(0) = $ 'Huey', $f(1) = $ 'Dewey', $f(2) = $ 'Louie', $g($'Huey'$) = 0$, $g($'Dewey'$) = 1$ and $g($'Louie'$) = 2$.

An injective mapping from a set with three elements to a set with four elements can be given, but not vice versa. For finite sets, according to this definition, two sets are equinumerous if they have the same number of elements, exactly what is intended.

But what about infinite sets? We have already seen that $\mathbb{N} \preceq \mathbb{N}^+$, and because of $\mathbb{N}^+ \subseteq \mathbb{N}$ holds also $\mathbb{N}^+ \preceq \mathbb{N}$, so also $\mathbb{N}^+ \simeq \mathbb{N}$. So the distinct sets $\mathbb{N}^+$ and $\mathbb{N}$ are equinumerous.

We will now show that $\mathbb{N} \simeq \mathbb{Z}$. One direction $\mathbb{N} \preceq \mathbb{Z}$ holds because of $\mathbb{N} \subseteq \mathbb{Z}$. To show the other direction $\mathbb{Z} \preceq \mathbb{N}$ we give an injective map $f : \mathbb{Z} \to \mathbb{N}$ as follows:

$$f(0) = 0, \ f(1) = 1, \ f(-1) = 2, \ f(2) = 3, \ f(-2) = 4, \ f(3) = 5, \ldots$$

So in general $f(n) = 2n - 1$ for every n in $\mathbb{N}^+$ and $f(-n) = 2n$ for every n in $\mathbb{N}$. In this way $f(x)$ is defined for every x in $\mathbb{Z}$. The function f is injective since no two distinct integers map to the same natural number, as is easily checked. Hence, we have proven that $\mathbb{N} \simeq \mathbb{Z}$.

In this construction we have not only created a map $f : \mathbb{Z} \to \mathbb{N}$ that is injective, but also has the additional property that every element of $\mathbb{N}$ is reached. This is called *surjective*. More precisely, a function $f : A \to B$ is called *surjective* if for every b in B some a in A exists such that $f(a) = b$.

A function $f : A \to B$ is called *bijective* if it is both injective and surjective. Let $f : A \to B$ be bijective. Then for every b in B there is at most one a in A with $f(a) = b$ since f is injective, but also at least one because f is surjective. So for every b in B there is *exactly one* a in A with $f(a) = b$. This means that there is also an function $g : B \to A$ defined by $g(b) = a$ if $b = f(a)$. This function g has the property that $g(f(a)) = a$ for every a in A, and $f(g(b)) = b$ for every b in B; it is called the *inverse* of f. For the example of the injective function $f : \mathbb{Z} \to \mathbb{N}$ that we gave, this function is bijective too, and for its inverse g we have

$$g(0) = 0, \ g(1) = 1, \ g(2) = -1, \ g(3) = 2, \ g(4) = -2, \ g(5) = 3, \ldots$$

In general: a bijective map always has an inverse, and that inverse is itself bijective.

In studying equivalence of sets, the following theorem of Schröder and Bernstein is a key property. It is called after Felix Bernstein (1878–1956) and Ernst Schröder (1841–1902); we give no proof here.

Theorem 4.2 *If there are injective functions $f : A \to B$ and $g : B \to A$, then also a bijective function from A to B exists.*

So two sets A and B are equinumerous if and only if a bijective function $f : A \to B$ exists.

Countable sets

A set A is called *countable* or *enumerable* if $A \preceq \mathbb{N}$. Finite sets are countable, and an infinite set is countable if it is equinumerous to $\mathbb{N}$. This means that there exists a bijective function $f : \mathbb{N} \to A$, so that by $f(0), f(1), f(2), \ldots$ exactly all elements of A are obtained: they are all different, but every element of A is in it. So the notion *countable* coincides with its suggested meaning that all elements of the set may be counted. The corresponding bijective function f is called the *enumeration*.

We have already seen that $\mathbb{N} \simeq \mathbb{Z}$, so that $\mathbb{Z}$ is countable. But what about the following number sets $\mathbb{Q}$, $\mathbb{R}$ and $\mathbb{C}$?

We will show that the set $\mathbb{Q}$ of rational numbers is countable. One way we to do so is by specifying an enumeration directly. But we will show more often in this book that an infinite set is countable. It appears to be useful to apply a general theorem for doing so. We now give that theorem first, and then use it to prove that $\mathbb{Q}$ is countable, and later also for other sets.

Theorem 4.3 *Assume that a function $g : A \to \mathbb{N}$ has the property that for every n in $\mathbb{N}$ only finitely many elements a in A exist such that $g(a) = n$. Then A is countable.*

Proof: If A is finite, then A is countable, so it remains to prove the theorem for A being infinite. We will do that by constructing a function $f : \mathbb{N} \to A$ that is bijective. First we look at all elements a of A for which $g(a) = 0$. There are finitely many of them, let's say k_0, and we number them as $a_0, a_1, \ldots, a_{k_0-1}$. Next we look at all elements a of A for which $g(a) = 1$. Again there are finitely many of them, let's say k_1, and we number them as $a_{k_0}, a_{k_0+1}, \ldots, a_{k_0+k_1-1}$. Similarly, the k_2 elements of A with $g(a) = 2$ are numbered starting in $a_{k_0+k_1}$. We continue this for all n in $\mathbb{N}$, of which there are always a finite number, k_n, with $g(a) = n$.

Now we define $f(n) = a_n$ for every n in $\mathbb{N}$. This is indeed a function $f : \mathbb{N} \to A$. Because all a_n are distinct, this function is injective. And because every element of A occurs as a_n for some n, this function is surjective. This shows that $f : \mathbb{N} \to A$ is a bijective mapping, hence A is countable. $\square$

As the first example of applying Theorem 4.3, let's once again prove that $\mathbb{Z}$ is countable. We define $g : \mathbb{Z} \to \mathbb{N}$ by defining $g(x) = |x|$ for all x in $\mathbb{Z}$. Here is $|x|$ the absolute value of x: for values $x \geq 0$ it is x, and for values $x < 0$ it is $-x$. Now $x = 0$ is the only element of $\mathbb{Z}$ for which $|x| = 0$. So there is exactly one element x with $g(x) = 0$, being a finite number of elements. For all other n in $\mathbb{N}$ there are exactly two elements of $\mathbb{Z}$ for which $|x| = n$, namely n and $-n$, again being a finite number of elements. Since this satisfies the condition of Theorem 4.3, it proves that $\mathbb{Z}$ is countable.

Now we are ready to prove the promised property for $\mathbb{Q}$.

Theorem 4.4 *The set $\mathbb{Q}$ of rational numbers is countable.*

Proof: Every rational number q can be written as $\frac{x}{m}$ with x in $\mathbb{Z}$ and m in $\mathbb{N}^+$. We define $g(q)$ as the smallest natural number k for which q can be written as $\frac{x}{m}$ with $|x| \le k$ and $m \le k$. For example $g(-23) = g(\frac{-23}{1}) = 23$, $g(\frac{3}{89}) = 89$ and $g(\frac{3}{87}) = g(\frac{1}{29}) = 29$. We observe that for every n in $\mathbb{N}$ there only finitely many elements q in $\mathbb{Q}$ exist satisfying $g(q) \le n$ exist, since all of them can be written as $\frac{x}{m}$ with $-n \le x \le n$ and $0 < m \le n$ for which there are only finitely many. So this function g satisfies the condition of Theorem 4.3, proving that $\mathbb{Q}$ is countable. $\square$

We now mention a few more general properties of countability, all of which are easy to prove, often using Theorem 4.3.

- Every subset of a countable set is countable.
- If $\sim$ is an equivalence relation on a countable set A, then $A/\sim$ is also countable.
- If $f : A \to B$ is an injective map and B is countable, then A is also countable.
- If $f : A \to B$ is a surjective map and A is countable, then B is also countable.
- If A and B are countable sets, then the Cartesian product $A \times B$ is also countable.

A proof using Theorem 4.3 is convincing in itself, but for some people it may help to see a picture of the enumeration. We now give a picture of an enumeration of $\mathbb{N} \times \mathbb{N}$ using a path starting in $(0,0)$:

$$
\begin{array}{ccccc}
\vdots & \vdots & \vdots & \vdots & \vdots \\
(0,3) & (1,3) & (2,3) & (3,3) & (4,3) \quad \cdots \\
(0,2) & (1,2) & (2,2) & (3,2) & (4,2) \quad \cdots \\
(0,1) & (1,1) & (2,1) & (3,1) & (4,1) \quad \cdots \\
(0,0) & (1,0) & (2,0) & (3,0) & (4,0) \quad \cdots
\end{array}
$$

The intended enumeration is obtained by following the indicated path from (0.0):

$$f(0) = (0,0), \ f(1) = (0,1), \ f(2) = (1,1), \ f(3) = (1,0), \ f(4) = (2,0),$$
$$f(5) = (2,1), \ f(6) = (2,2), \ f(7) = (2,3), \ f(8) = (3,3), \ \ldots$$

Since all elements of $\mathbb{N} \times \mathbb{N}$ occur on this path, indeed the path defines an enumeration, proving that $\mathbb{N} \times \mathbb{N}$ is countable.

Applying the general properties of countability may be repeated. For instance, the construction of the Cartesian product may be repeated any finite number of times. As an example, we conclude that the set of all five tuples (n, q, m, k, r) where n is a natural number, m and k are integers, and q and r are rational numbers, is countable.

This may suggest that all infinite sets are countable, but that is not the case. In the next section we will see why.

Countability roughly means: no more than the natural numbers. With this focus on natural numbers let's return now to the challenge in the chapter on natural numbers. It was about the numbers with exactly 109 divisors, excluding the number itself and the number 1. If we do include the number itself and the number 1, there are exactly 111 divisors, and that is 3×37, in which 3 and 37 are prime numbers. We have seen that any natural number n can be written as

$$n = p_1^{m_1} \times p_2^{m_2} \times \cdots \times p_k^{m_k},$$

where $p_1 < p_2 < \cdots < p_k$ are prime numbers, this was Theorem 2.3. Now we determine what the divisors of this are: these are exactly the numbers of the form $p_1^{i_1} \times p_2^{i_2} \times \cdots \times p_k^{i_k}$ with $0 \leq i_j \leq m_j$ for every j from 1 through k. If all i_j's are equal to 0 we get the divisor 1 (recall $p^0 = 1$), and if every i_j is equal to m_j we get the divisor n. For i_1 we have $m_1 + 1$ possibilities, namely all numbers from 0 to m_1. Similarly we have $m_j + 1$ possibilities for i_j, for every j. So the total number of divisors, including 1 and n itself, is exactly

$$(m_1 + 1) \times (m_2 + 1) \times \cdots \times (m_k + 1).$$

If the total number of divisors has to be exactly $111 = 3 \times 37$, this only works if $k = 2$, and $m_1 = 36$ and $m_2 = 2$, or vice versa: $m_1 = 2$ and $m_2 = 36$. This is exactly true for all numbers $n = p^{36} \times q^2$ for which p and q are two distinct prime numbers. Because there are infinitely many prime numbers (Theorem 2.1) this can be done in infinitely many ways, and the smallest n for which this holds is $n = 2^{36} \times 3^2 = 618475290624$, and that is indeed just over 600 billion.

This conclusion can be drawn without using a calculator as follows. From $2^{10} = 1024$ it follows that $2^{30} = (1024)^3$, being a number just over 1.06×10^9. To get 2^{36} we multiply this by $2^6 = 64$, yielding more than 6.8×10^{10}. And we have to multiply that by $3^2 = 9$, yielding more than $6 \times 10^{11} = 600$ billion. Anyway, the challenge about the number of divisors has been solved.

Uncountable sets

A set is called *uncountable* if it is not countable. We will now show that the set $\mathbb{R}$ of real numbers is uncountable. To do so, we use the usual decimal notation of real numbers. Suppose $\mathbb{R}$ is countable. Then also every subset of $\mathbb{R}$ countable. Our goal is to derive a contradiction. We consider the subset A of $\mathbb{R}$ consisting of all real numbers ≥ 0 and < 1 that have only zeros and ones in their decimal notation. Numbers with finite decimal notation, such as 0.1 we think to be continued by infinitely many zeros, so all numbers a in A can be written as 0. followed by an infinite sequence consisting of the digits 0 and 1:

$$a = 0.a_0 a_1 a_2 a_3 \cdots .$$

Due to the assumption that the infinite set A is countable, there is a bijective function $f : \mathbb{N} \to A$. For every n we write $f(n)$ in decimal notation:

$$f(n) = 0.f(n)_0 f(n)_1 f(n)_2 f(n)_3 \cdots ,$$

where every $f(n)_m$ equals 0 or 1.

Now we construct a special number

$$b = 0.b_0 b_1 b_2 b_3 \cdots$$

For every n in $\mathbb{N}$, we define $b_n = 0$ if $f(n)_n = 1$, and $b_n = 1$ if $f(n)_n = 0$ Since every b_n is equal to 0 or 1, the number b is in A. Since f is bijective, some n exists such that $f(n) = b$. For this particular n we consider b_n. If $b_n = 0$, then $b_n = f(n)_n = 1$. And if $b_n = 1$, then $b_n = f(n)_n = 0$. In both cases we have a contradiction, so this produces the desired contradiction. So we have shown that the original assumption that $\mathbb{R}$ is countable, is not true. This proves the following:

Theorem 4.5 *The $\mathbb{R}$ set of real numbers is uncountable.*

The argument we gave for this is called the *diagonal argument*. Indeed, we have constructed the sequence $b_0, b_1, b_2, \ldots$ and thus the number b by flipping the underlined numbers in the following picture. Here, every row forms the decimal notation of some number $f(n)$, and these underlined numbers exactly form the diagonal:

$$
\begin{array}{ccccc}
\underline{f(0)_0} & f(0)_1 & f(0)_2 & f(0)_3 & f(0)_4 & \cdots \\
f(1)_0 & \underline{f(1)_1} & f(1)_2 & f(1)_3 & f(1)_4 & \cdots \\
f(2)_0 & f(2)_1 & \underline{f(2)_2} & f(2)_3 & f(2)_4 & \cdots \\
f(3)_0 & f(3)_1 & f(3)_2 & \underline{f(3)_3} & f(3)_4 & \cdots \\
f(4)_0 & f(4)_1 & f(4)_2 & f(4)_3 & \underline{f(4)_4} & \cdots \\
\vdots & \vdots & \vdots & \vdots & \vdots
\end{array}
$$

Since $\mathbb{R}$ is a subset of $\mathbb{C}$ it follows from Theorem 4.5 that also the set $\mathbb{C}$ of complex numbers is uncountable.

This uncountability result, and the diagonal argument to prove it, originates from the German mathematician Georg Cantor (1845–1918). By then it was a quite unusual way of reasoning, by which the result was somewhat controversial to some people. It turns out to be a powerful means of demonstrating the existence of all kinds of objects in a non-constructive way. Next we will show how to use it to prove that incomputable numbers exist.

Computable numbers

We call a real number x *computable* if there is a (finite) computer program that produces the decimal representation of x when executed. It does not really matter in which programming language that program is written, since there are programs that convert a program in one language into a program in another language with the same output. Here we focus on the main ideas and omit a more precise definition of of computable. A computer program in which we abstract from the details of the programming language is also called an *algorithm*. It also doesn't matter whether the usual decimal or binary representation of numbers is used, or some other representation, as long as these representations can be transformed to each other by some algorithm. In fact, we should say 'a decimal representation' rather than 'the decimal representation', because, for example 1 and $0.99999\cdots$ both represent the same number 1, but that is also irrelevant for this definition.

If we assume that our programming language has a command `write` for writing output, then the program

```
write('23')
```

shows that the number 23 is computable. Assuming that our programming language has a common `while` construction, the program

```
write('0.'); while true do write('3')
```

returns as output

$$0.3333333\cdots,$$

that is, the decimal representation for $\frac{1}{3}$, showing that $\frac{1}{3}$ is computable. It is more complicated to show that $\sqrt{2}$ and π are also computable, but this can be done by constructing algorithms that yield the infinite decimal representation of those numbers.

But are there also real numbers that are not computable? Yes, that follows from the following proposition.

Theorem 4.6 *The set of all computable real numbers is countable.*

Proof: We will use Theorem 4.3 again. Fix some programming language. For any computable number x, we define $g(x)$ as the smallest number n for which there is a program in this language consisting of n symbols that has the decimal representation of x as output. We observe that for every n there are only finitely many programs that can be represented in n symbols: if the programs are expressed in the 26 letters of the alphabet, there are at most 26^n, and in fact much less because not every sequence of n letters is a correct program. Anyway, the condition of Theorem 4.3 holds, and we conclude that the set of all computable real numbers is countable. $\square$

Since the set of all computable real numbers is countable and the $\mathbb{R}$ set of all real numbers is uncountable, we conclude that non-computable numbers exist: even very many, uncountably many. However, we cannot construct one, because as soon as we can, that number is computable. Due to these counterintuitive observations, it is not surprising that Cantor's insights were initially controversial.

Cardinal numbers

What is the number of elements in a set? We may define these numbers as equivalence classes on sets, where two sets A and B are equivalent if there is a bijective mapping between them, i.e. $A \simeq B$. For finite sets this yields the following equivalence classes :

- the class consisting of the empty set, with 0 elements,
- the class consisting of all sets with one single element,

- the class consisting of all two-element sets,
- the class consisting of all three-element sets,
- the class consisting of all four-element sets,
- ...

exactly corresponding to the ordinary natural numbers. Therefore these classes are denoted by $0, 1, 2, 3, \ldots$, just like the natural numbers themselves. This may even be considered as a definition of natural numbers. But these equivalence classes also exist for infinite sets. The equivalence classes are called *cardinal numbers*. The finite cardinal numbers coincide with the natural numbers, and the smallest infinite cardinal number is the class of infinite countable sets. It is denoted by $\aleph_0$, pronounced as *alef zero*. The symbol $\aleph$ is the *Aleph*, the first letter of the Hebrew alphabet. Greek letters are often used in mathematics, but many Greek letters already have specific meanings, such as the ϵ in the definition of limit, so this notation switched to the next classical language with its own alphabet.

Just as $x < y$ holds exactly if $x \leq y$ holds and $y \leq x$ does not hold, we also define $A \prec B$ for sets A, B if and only if $A \preceq B$ holds and $B \preceq A$ does not hold. We can see A as a strictly smaller set than B, for instance we conclude $\mathbb{N} \prec \mathbb{R}$ from Theorem 4.5.

Apart from the cardinal number $\aleph_0$ of $\mathbb{N}$ and the cardinal number of $\mathbb{R}$, are there any other infinite cardinal numbers? Indeed there are. By a diagonal argument similar to Theorem 4.5 one shows that for every set A it holds that

$$A \prec \mathcal{P}(A),$$

where $\mathcal{P}(A)$ denotes the set of all subsets of A, also called *power set* of A. That gives rise to infinitely many different infinite cardinal numbers: from

$$\mathbb{N} \prec \mathcal{P}(\mathbb{N}) \prec \mathcal{P}(\mathcal{P}(\mathbb{N})) \prec \mathcal{P}(\mathcal{P}(\mathcal{P}(\mathbb{N}))) \prec \cdots$$

follows that the cardinal numbers of the infinite sets $\mathbb{N}$, $\mathcal{P}(\mathbb{N})$, $\mathcal{P}(\mathcal{P}(\mathbb{N}))$, $\mathcal{P}(\mathcal{P}(\mathcal{P}(\mathbb{N})))$, ... are all distinct. So there are infinitely many different kinds of infinity!

The $\aleph_0$ notation suggests that there are also things like $\aleph_1$ and $\aleph_2$. Indeed, $\aleph_1$ is defined as the smallest cardinal number strictly greater than $\aleph_0$, and $\aleph_2$ is defined as the smallest cardinal number strictly greater than $\aleph_1$, and so on. It has been proven that these cardinal numbers all exist, assuming *axiom of choice*, but we do not go into further detail here.

One can show that $\mathcal{P}(\mathbb{N}) \simeq \mathbb{R}$. This suggests that the cardinal number of $\mathbb{R}$ may be $\aleph_1$. From $\mathbb{R}$ we know that $\mathbb{N} \prec \mathbb{R}$. If there were another set A with $\mathbb{N} \prec A \prec \mathbb{R}$, so in between $\mathbb{N}$ and $\mathbb{R}$, then the cardinal number of $\mathbb{R}$ is not $\aleph_1$. If no such A exists then the cardinal of $\mathbb{R}$ is $\aleph_1$. This latter option that the cardinal number of $\mathbb{R}$ equals $\aleph_1$ is called the *continuum hypothesis*. It is not clear whether this is true or not. It has even been shown that this statement cannot be deduced from, and is therefore independent of, a common set of axioms for set theory, established by Ernst Zermelo (1871–1953) and Abraham Fraenkel (1891–1965).

Challenge: monotone functions

In this chapter, functions between infinite sets played an key role, so it is appropriate to have a challenge about that.

A function $f : \mathbb{N} \to \mathbb{N}$ is called *monotone* if $f(n) > f(m)$ holds for every $n > m$.

For example, let f and g be defined by $f(n) = 3n$ and $g(n) = n + 1$ for every n in $\mathbb{N}$. Then both f and g are monotone. For these f, g, and for every n in $\mathbb{N}$ it holds

$$g(g(f(n))) = 3n + 2 < 3n + 3 = f(g(n)).$$

So not a single n in $\mathbb{N}$ exists such that $g(g(f(n))) \geq f(g(n))$. In the challenge we consider a variant of this property.

Challenge:

Let $f, g : \mathbb{N} \to \mathbb{N}$ both be monotone. Is it always the case that some n in $\mathbb{N}$ exists such that $g(f(f(n))) \geq f(g(n))$?

---

5 Infinite sequences

In previous chapters, we have seen several infinite sets of numbers: natural, integer, rational, real and complex numbers. The first three types of numbers, natural, integer and rational numbers, could all be represented finitely. For real numbers, this is not possible anymore since there are uncountably many of them. One standard way to represent real numbers infinitely is by some integer in front of a decimal point, followed by an infinite sequence of digits behind the decimal point. Now we will look at the simplest form of objects that can no longer be finitely represented themselves: *infinite sequences*. For this, we need an *alphabet*: the set of elements that may occur in the sequence. In many cases, the alphabet A only consists of two elements 0 and 1. For example, we may think of the sequence of decimal digits after the decimal point of a real number as an infinite sequence over the alphabet consisting of the digits 0 through 9.

In general, an infinite sequence a over an alphabet A is given as

$$a = a_0, a_1, a_2, a_3, a_4, \ldots$$

where a_n is an element of A for every natural number n. We will often omit the commas in this notation from now on, very useful if we want to show a large starting piece of a specific sequence. It is good to realize that such a sequence is the same as a function $a : \mathbb{N} \to A$, where you write a_n instead of $a(n)$.

The set of all infinite sequences over an alphabet A is denoted by A^∞. Here ∞ is the usual symbol for infinity; this symbol is also known as *lemniscate*, a word that comes from Greek via Latin and means *band* or *ribbon*.

If A contains at least two elements, then A^∞ is uncountable. In our proof with the diagonal argument that $\mathbb{R}$ is uncountable, Theorem 4.5, we actually

DOI: 10.1201/9781003466000-5

proved that the set of infinite sequences over the alphabet containing only 0 and 1 is uncountable. Exactly the same proof holds if the alphabet contains more than two elements.

For a finite alphabet A, we want to investigate what kinds of infinite sequences over A can be made. In this chapter, we focus on the simplest: *periodic* and *ultimately periodic* sequences. In a later chapter, we will consider *morphic* sequences, being the basis of most turtle figures we will present in this book. But first we describe some basic operations on infinite sequences, just as we had basic operations like successor, addition and multiplication on natural numbers.

Operations on sequences

Since we will be talking about infinite sequences from now on, we will call them *sequences* for short. In order not to get confused with finite sequences of symbols, we will call these finite sequences *words* from now on, just as it is customary for a word to consist of a finite sequence of letters. The set of sequences over an alphabet A we already denoted by A^∞. The set words over A is denoted by A^*. A word may be empty, or may consist of one single symbol, or any finite number of symbols. The empty word is indicated by the Greek letter ϵ (epsilon), say, the Greek e, and it is no coincidence that that is the first letter of the word *empty*. In older texts, the Greek letter λ (lambda) is also used, say, the Greek ℓ, being the first letter of the German word *leer* for *empty*.

A set of words is called a *(formal) language*. The structure of such formal languages has been investigated extensively. As this is outside the focus of this book, we will not go into here. Almost every university computer science curriculum has a compulsory course on formal languages, and I even taught such a course myself several times.

Back to (infinite) sequences. We want to investigate some basic operations on this. The first one describes removing the first element from a sequence, leaving the *tail* of that sequence. This operation is denoted by tail, and is defined by $\text{tail}(a)_i = a_{i+1}$ for every natural number i. So for $a = a_0 a_1 a_2 a_3 a_4 \cdots$ we have

$$\text{tail}(a) = a_1 a_2 a_3 a_4 \cdots$$

Instead of just removing the first element, we may also remove a lot of elements. There are many ways to do this. An interesting one is even, defined by $\text{even}(a)_i = a_{2i}$ for every natural number i. So for $a = a_0a_1a_2a_3a_4 \cdots$ we have

$$\text{even}(a) = a_0a_2a_4a_6 \cdots$$

In a similar way also odd is defined, note that $\text{odd}(a)$ is the same as $\text{even}(\text{tail}(a))$.

Just as we have general properties such as commutativity and associativity for addition and multiplication, we have general properties for operations on sequences. A first one reads: every sequence a satisfies

$$\text{tail}(\text{even}(a)) = \text{even}(\text{tail}(\text{tail}(a))).$$

This is easily deduced from the definitions of tail and even: for every i, we have

$$\text{tail}(\text{even}(a))_i = \text{even}(a)_{i+1} = a_{2(i+1)} = a_{2i+1+1} =$$

$$\text{tail}(\text{tail}(a))_{2i} = \text{even}(\text{tail}(\text{tail}(a)))_i.$$

The operations tail and even only remove elements, but also elements may be added or changed. If a is a sequence over A, and x is an element of A, the sequence xa is obtained by putting x in front. So we have

$$(xa)_0 = x \text{ and } (xa)_i = a_{i-1} \text{ for each } i > 0.$$

It follows immediately that $\text{tail}(xa) = a$ for every a, x, since for every $i \geq 0$ we have $\text{tail}(xa)_i = (xa)_{i+1} = a_i$.

Instead of only a single symbol, also any word may be added in front of a sequence. If a is a sequence over A and $u = u_0u_1 \ldots u_{n-1}$ is a word of length n over A, then ua is the sequence obtained by putting u in front of a, so $(ua)_i = u_i$ if $i < n$, and $(ua)_i = a_{i-n}$ if $i \geq n$.

Morphisms

So far we've only seen operations on sequences that remove or add elements, but don't change them. An important operation to change elements is based on the notion *morphism*. A *morphism* is defined to be a function $f : A \to A^*$. So a morphism f replaces every symbol x from an alphabet A by a word $f(x)$ in A^*. Such a morphism is not only applied to individual symbols but also to words and sequences. The idea is that every symbol x in the word or sequence is replaced by $f(x)$. For a sequence $a = a_0a_1a_2a_3a_4 \ldots$ that means that the sequence $f(a)$ is defined by

$$f(a) = f(a_0)f(\text{tail}(a)).$$

To give an example: if A consists of 0 and 1, and $f(0) = 10$ and $f(1) = 001$, and $a = 001011\ldots$, then

$$f(a) = f(0)f(0)f(1)f(0)f(1)f(1) \cdots = 101000110001001 \cdots$$

Does a morphism applied to an infinite sequence always produce an infinite sequence? That turns out not to be the case. The empty word ϵ is also in A^*, so it is allowed that $f(x) = \epsilon$, and that may cause that the result is not an infinite sequence. For example, when defining that $f(x) = \epsilon$ for every x in A, then by applying f on any sequence replaces every symbol by the empty word, resulting in no symbol at all, being the empty word, which is certainly not an infinite sequence.

To avoid these problems, we only allow morphisms $f : A \rightarrow A^+$, where A^+ is defined as the set of non-empty words over A. Then in the construction of $f(a)$ from a, each symbol x is replaced by the non-empty word $f(x)$, and the result is always an infinite sequence.

These morphisms will later play a crucial role in the definition of *morphic sequences* of which we will study all kinds of turtle figures. But first we look at simpler sequences.

Periodic and ultimately periodic sequences

The simplest infinite sequence we can imagine consists only of zeros:

$$\text{zeros} = 0000000 \cdots$$

Still very simple is the alternating sequence

$$\text{alt} = 010101010 \cdots,$$

consisting of alternating 0 and 1.

More generally, choosing a non-empty *word* u over A, being a finite sequence of symbols from A, defines the sequence

$$u^\infty = uuuuu \cdots$$

So using this notation we have $\text{zeros} = (0)^\infty$ and $\text{alt} = (01)^\infty$. In general, if 0 is in A and the morphism $f : A \rightarrow A^+$ satisfies $f(0) = u$, then $u^\infty = f(\text{zeros})$.

We say that an infinite sequence a over A *periodic* if there is a non-empty word u over A such that $a = u^\infty$. The length of the shortest u for which this applies is then called the *period* of the periodic sequence. So our sequence zeros $= 0^\infty = (00)^\infty = (000)^\infty$ is periodic with period 1, and alt $= (01)^\infty$ is periodic with period 2.

The following theorem gives some basic properties of periodic sequences.

Theorem 5.1 *If a is a periodic sequence, then so are* tail(a) *and* even(a).

If a is a periodic sequence over A and $f : A \to A^+$ *is a morphism, then* $f(a)$ *is also a periodic sequence.*

Proof: (sketch)

If a is a periodic sequence over an alphabet A, then there is a non-empty word u such that $a = u^\infty$. Write $u = u_0 u_1 \ldots u_{n-1}$ where $u_0, u_1, \ldots, u_{n-1}$ are in A.

Then tail$(a) = (u_1 u_2 \ldots u_{n-1} u_0)^\infty$.

If n is even, then even$(a) = (u_0 u_2 \ldots u_{n-2})^\infty$. If n is odd, then even$(a) = (u_0 u_2 \ldots u_{n-1} u_1 u_3 \ldots u_{n-2})^\infty$.

If $f : A \to A^+$, then $f(a) = (f(u_0) f(u_1) \ldots f(u_{n-1}))^\infty$. $\square$

Not all basic operations on sequences applied on a periodic sequence always yield a periodic sequence again. In particular, this occurs for adding a word or symbol in front of a sequence. For example: zeros is periodic, but

$$1\text{zeros} = 10000000 \cdots$$

is not periodic.

Now we define an extension of the term periodic sequence. We call a sequence a over an alphabet A *ultimately periodic* if there is a word v in A^* is and a non-empty word u in A^+ such that $a = v u^\infty$.

It is clear that every periodic sequence is ultimately periodic, take for example $v = \epsilon$, but the reverse is not true: 1zeros is ultimately periodic but not periodic.

It is easy to see that if we put a word w in front of any ultimately periodic sequence $a = vu^\infty$, the result is again ultimately periodic: $wa = wvu^\infty$, so wa is obtained by putting the word wv in front of the periodic sequence u^∞.

And also for all the other operations we have looked at, a similar property holds: if we apply tail or even or a morphism to an ultimately periodic sequence, then the result is again an ultimately periodic sequence. We say that the ultimately periodic sequences are *closed* under the operations tail, even, morphisms, and adding a word in front, just as the natural numbers are closed under the operations successor, addition and multiplication.

Decimal notation of numbers

For a real number r satisfying $0 \le r < 1$, we have its standard decimal notation: it starts with '0.', and is then followed by an infinite sequence a over the alphabet consisting of the digits 0 through 9. Well-known is

$$\frac{1}{3} = 0.3333333\cdots,$$

where $a = 3^\infty$ consists only of the digit 3. A finite decimal notation may be extended by zeros to fit in this format, for instance

$$\frac{1}{4} = 0.25 = 0.25000000\cdots.$$

One may also write

$$\frac{1}{4} = 0.24999999\cdots,$$

but we prefer the first notation since then the finite notation is just obtained by ignoring the infinite sequence of zeros. Two more examples:

$$\frac{1}{7} = 0.142857142857142857\cdots \quad \text{and} \quad \frac{893}{3700} = 0.24135135135\cdots,$$

with $a = (142857)^\infty$, and $a = 24(135)^\infty$ respectively.

From these examples we observe that for *rational numbers* r the decimal notation of r is of the shape $0.a$ where a is an ultimately periodic sequence. Is this a coincidence, or is it always like this? We will now show that it always holds, even stronger, we have the following theorem:

Theorem 5.2 *A real number r with $0 \le r < 1$ is rational if and only if in its decimal notation $0.a$ the sequence a is ultimately periodic.*

Before we give the proof, we give some general remarks. We discuss here the *decimal* notation, also called the notation in base 10, using the 10 digits from 0 to 9. In the arguments we give, this number base 10 does not play a role at all; it also applies to the notation in base n for every $n > 1$, in particular the *binary* notation in base 2, only using the digits 0 and 1. The reason that decimal has become our standard stems from the fact that most people have ten fingers, and counting on the fingers comes naturally to us.

Another comment is about what kind of statement this is. We have given our framework of numbers as natural as possible, from simple to more diffi-cult: starting by natural numbers, then extending to integers, then to rational numbers and then to real numbers. Quite apart from that, in this chapter, we started building infinite sequences from simple to more difficult: from periodic sequences to ultimately periodic sequences, and in Chapter 6, we will extend this once more to morphic sequences. Now it turns out that for real numbers there is a way to represent them using infinite sequences. For general real numbers, the part in front of the decimal point is more general, but restricting to real numbers between 0 and 1 it is just '0.' followed by and infinite sequence of symbols. And now Theorem 5.2 states that the rational numbers, forming a natural subclass of the real numbers, correspond exactly to the ultimate periodic sequences, a natural class of infinite sequences. This strong correspondence now confirms that the rational numbers and the ultimately periodic sequences are natural subclasses of the real numbers and the infinite sequences, respectively. Mathematicians like to see this kind of relationship.

But now it is time to give the proof of Theorem 5.2.

Proof: Take any real number r with $0 \le r < 1$ having decimal notation $0.a$. For proving the theorem we have to show two directions. First we show that if a is ultimately periodic, then r is a rational number. We assume that a is ultimately periodic, that is, we write $a = v\,u^\infty$ where v and u are words over the alphabet from 0 through 9, and where u is not empty. Let $n > 0$ be the length of u. Let's now look at the decimal notation of $\frac{1}{10^n-1}$. For $n = 1$ we have

$$\frac{1}{10 - 1} = \frac{1}{9} = 0.1111111\cdots$$

For $n = 2$ we have

$$\frac{1}{10^2 - 1} = \frac{1}{99} = 0.01010101\cdots$$

For $n = 3$ we have

$$\frac{1}{10^3 - 1} = \frac{1}{999} = 0.001001001001\cdots$$

This pattern continues: for every $n > 0$, the decimal notation of $\frac{1}{10^n-1}$ reads $0.(0^{n-1}1)^\infty$. Now let n_u be the natural number represented by the word u. Then we obtain

$$\frac{n_u}{10^n - 1} = n_u \times 0.(0^{n-1}1)^\infty = 0.u^\infty.$$

An example of this is $u = 347$, then we get

$$\frac{347}{999} = 0.347347347347\cdots$$

So we have shown that $0.u^\infty$ is a rational number. If v is empty, we're done. If v is not empty and $m > 0$ is the length of v, then $0.v$ is the decimal notation of $\frac{n_v}{10^m}$, where n_v is the number is indicated by the word v. If we add $\frac{n_u}{10^m(10^n-1)}$ to this, i.e. the number represented by $0.0^m u^\infty$, we get exactly the number represented by $0.a = 0.v\,u^\infty$. Hence the number r having decimal notation $0.a$ is the rational number

$$\frac{n_v}{10^m} + \frac{n_u}{10^m(10^n-1)} = \frac{n_v(10^n-1)+n_u}{10^m(10^n-1)}.$$

To give an example: $0.2361(347)^\infty$ can be written as the rational number

$$\frac{2361 \times 999 + 347}{9,990,000}.$$

This proves one direction of the theorem.

For the other direction, we assume that r is any rational number with $0 \le r < 1$ whose decimal notation is $0.a$, and now we have to show that a is ultimately periodic. We write r as $r = \frac{m}{n}$ where $0 \le m < n$ are natural numbers. If c is the first digit of a, then the decimal notation of $r - \frac{c}{10}$ consists of a 0 after the decimal point followed by tail(a). If we multiply this number by 10, its decimal notation is obtained by removing the 0 after the decimal point. This yields $0.\mathrm{tail}(a)$, representing the number $10(r - \frac{c}{10}) = 10r - c$. Since $r = \frac{m}{n}$ this can be written as $\frac{10m-cn}{n}$, so it is of the form $\frac{m'}{n}$ for some natural number $m' < n$. This argument is repeated: also the number with decimal notation $0.\mathrm{tail}(\mathrm{tail}(a))$ is of the form $\frac{m''}{n}$ for some natural number $m'' < n$. Writing

$$\mathrm{tail}^k(a) = \underbrace{\mathrm{tail}(\mathrm{tail}(\cdots \mathrm{tail}(a)\cdots)),}_{k \text{ times } \mathrm{tail}}$$

then also for every k the number with decimal notation $0.\mathrm{tail}^k(a)$ is of the form $\frac{m_k}{n}$ for $m_k < n$. Now we arrive at the key argument: there are finitely many natural numbers $m_k < n$, and there are infinitely many different ks, so for some k we obtain s value m_k that we had before. In other words: there exist $p < q$ with $m_p = m_q$. But that means $0.\mathrm{tail}^p(a)$ is the same number as $0.\mathrm{tail}^q(a)$. But then the sequence $\mathrm{tail}^p(a)$ must be equal to the sequence $\mathrm{tail}^q(a)$. Now let v be the word consisting of the first p symbols of a, and u be the word consisting of the following $q - p$ symbols, and b the remaining sequence. Then $a = vub$. It follows that $\mathrm{tail}^p(a) = ub$ and $\mathrm{tail}^q(a) = b$. But $\mathrm{tail}^p(a)$ and $\mathrm{tail}^q(a)$ were equal, so $ub = b$. From $ub = b$ it follows that

$b = u^\infty$, and so is $a = vub = vu^\infty$ ultimately periodic, exactly what we had to prove. $\square$

The last part of this argument may be interpreted as the observation that when performing the corresponding long division we end up in a pattern that keeps repeating itself.

Rational numbers also play an important role in the challenge on ten questions, so now we will give the solution for that. It is clear that many of the ten questions refer to each other, so it is useful to give a name to the answer to each of the ten questions. For $k = 1, 2, \ldots, 10$, let A_k be the answer to question k. For example, A_1 is the number of prime numbers < 20, which are $2, 3, 5, 7, 11, 13, 17, 19$. These are eight numbers in total, so $A_1 = 8$. The next one is A_2, which is the sum of all the others. Establishing A_2 has to be postponed until we have information about the other values. The same holds for A_3 and A_4, at this moment we only state

$$A_3 = \frac{A_8 + A_{10}}{2} - A_7 \text{ and } A_4 = \frac{A_6 + A_7}{10}.$$

Question 5 is about rational numbers and doesn't refer to any other questions, so we will determine A_5 now. It is defined as the number of distinct numbers of the shape $\frac{n}{m}$ where n, m are integers satisfying $-5 < n < 5$ and $0 < m < 10$. The positive numbers of these are

$$\frac{1}{m} \text{ for } m = 1, 2, \ldots, 9, \quad \frac{2}{m} \text{ for } m = 1, 3, 5, 7, 9,$$

$$\frac{3}{m} \text{ for } m = 1, 2, 4, 5, 7, 8, \text{ and } \frac{4}{m} \text{ for } m = 1, 3, 5, 7, 9,$$

that's $9 + 5 + 6 + 5 = 25$ in total. All these 25 have a negative counterpart. Together with the number 0 this yields 51 in total, so $A_5 = 51$.

The next two,

$$A_6 = \frac{A_4 + A_5}{3} \text{ and } A_7 = \frac{A_1 \times A_4}{2},$$

refer to others. This is not the case with the last three where we have to count the numbers x with $x^2 = n$ where n is an integer with $-10 < n < 10$. In question 8, these numbers must be rational, and we get three positive ones, namely $1, 2, 3$, the three negative counterparts of these, and the number 0, making 7 in total, so $A_8 = 7$. Question 9 concerns real numbers, then we have 9 positive, namely $\sqrt{1}, \sqrt{2}, \sqrt{3}, \ldots, \sqrt{9}$, 9 negative , and the number 0, yielding 19 in total, so $A_9 = 19$. Finally, question 10 asks for the number of such complex numbers. These are the 19 real numbers we already had, extended by the numbers $\sqrt{1}, \sqrt{2}, \sqrt{3}, \ldots, \sqrt{9}$ multiplied by i or $-i$, so 18 more numbers, yielding $A_{10} = 19 + 18 = 37$.

Now we come back to the remaining As that refer to each other for which we have

$$A_4 = \frac{A_6 + A_7}{10}, \ A_6 = \frac{A_4 + A_5}{3} = \frac{A_4 + 51}{3}, \ A_7 = \frac{A_1 \times A_4}{2} = \frac{8A_4}{2} = 4A_4.$$

If we substitute $A_7 = 4A_4$ in $10A_4 = A_6 + A_7$ we get $A_6 = 6A_4$. And using this in $3A_6 = A_4 + 51$ yields $18A_4 = A_4 + 51$. Hence $17A_4 = 51$, so $A_4 = 3$. Now using $A_7 = 4A_4$ and $A_6 = 6A_4$ gives $A_6 = 18$ and $A_7 = 12$. Next we obtain

$$A_3 = \frac{A_8 + A_{10}}{2} - A_7 = \frac{7 + 37}{2} - 12 = 10.$$

Now we are finished with all of them except A_2, and all these values except A_2 yield the sum $8 + 10 + 3 + 51 + 18 + 12 + 7 + 19 + 37 = 165$. Since A_2 equals this sum 165, the sum of all ten values equals $165 + 165 = 330$, which is the answer for this challenge. After having solved this challenge now, we return to the infinite sequences.

Frequency of symbols

The sequence zeros only contains zeros, so consists of 100 % zeros. If we divide this percentage by 100, we call it the *frequency* of the indicated symbol. This frequency of the symbol x in a sequence a is denoted by $[a]_x$. So we have $[\text{zeros}]_0 = 1$. In the sequence zeros there are no ones at all, so that gives a frequency of 0 of ones, denoted by $[\text{zeros}]_1 = 0$. The sequence alt consists half of zeros and half of ones, so $[\text{alt}]_0 = [\text{alt}]_1 = \frac{1}{2}$.

How to define this frequency $[a]_x$ exactly? For a finite word, the frequency of a symbol x is easily defined by dividing the number of occurrences of x in the word by the length of the word. Since a sequence is infinite, it becomes a bit trickier: both zeros and alt are infinitely long and contain infinitely many zeros. So then both $[\text{zeros}]_0$ and $[\text{alt}]_0$ would be infinity divided by infinity, while we want the result to be 1 for the one and $\frac{1}{2}$ for the other. Fortunately, in a previous chapter in the section on real numbers, we introduced the concept of *limit*, and that turns out to be exactly what we need here. Recall the definition of limit:

A sequence $r_0, r_1, r_2, r_3, r_4, \ldots$ of real numbers has *limit L* if for every $\epsilon > 0$ there exists N such that for every $n > N$ holds:

$$L - \epsilon < r_n < L + \epsilon.$$

Intuitively this states: the bigger n, the closer the number r_n gets to that limit L. Not every sequence has a limit, but if it has a limit L, then L is also the only number that meets this definition of limit.

In defining the frequency $[a]_x$ of a symbol x in a sequence a, we now avoid the complication of dividing infinity by infinity by not taking the entire sequence a, but for each n we take the word consisting of the first $n + 1$ symbols. Next we define r_n to be the frequency of the symbol x in that word. Let w_n be the word consisting of the first n symbols of the sequence a, among which there are $|w_n|_x$ symbols equal to x, then

$$r_n = \frac{|w_{n+1}|_x}{n+1}.$$

Here we take $n + 1$ in order to prevent r_0 becoming equal to $\frac{0}{0}$.

For large n, we can see r_n as a good estimate of the frequency $[a]_x$. We now define:

$$[a]_x = \text{The limit of this sequence } r_0, r_1, r_2, r_3, r_4 \cdots$$

We will use this notation $[a]_x$ only for infinite sequences a, and the notation $|w|_x$ only for finite words w.

Let's now compute $[a]_0$ defined in this way for a few sequences a we have seen. The sequence $a = \text{zeros}$ contains only zeros, so $r_n = 1$ for every n. And indeed $L = 1$ then satisfies the definition of limit, because $1 - \epsilon < 1 < 1 + \epsilon$ for every $\epsilon > 0$. So $[\text{zeros}]_0 = 1$, and likewise $[\text{zeros}]_1 = 0$ because there are no ones at all, so $r_n = 0$ for every n.

In the next example, we choose $a = 1\text{zeros} = 100000\cdots$, and $x = 0$. Now the first n symbols contain exactly $n - 1$ zeros, yielding

$$r_n = \frac{n-1}{n}$$

for every n. This sequence

$$0, \frac{1}{2}, \frac{2}{3}, \frac{3}{4}, \frac{4}{5}, \cdots$$

has limit 1, as can be seen as follows. For any $\epsilon > 0$ we choose a number n with $n > \frac{1}{\epsilon}$, then

$$1 - \epsilon < \frac{n-1}{n} < 1 + \epsilon.$$

So according to the definition of limit, we obtain $[1zeros]_0 = [zeros]_0 = 1$. In general, the effect of the first element on r_n decreases as n increases. This is not only the case when adding a single symbol to the front of the sequence, but also if you add any (finite) word to the front. Again exploiting the definition of limit, this yields the following:

Theorem 5.3 *If a is a sequence and x is a symbol for which $[a]_x$ exists, and w is any word, then $[wa]_x = [a]_x$.*

Let's consider one more example: $a = $ alt. We expect $[alt]_0 = \frac{1}{2}$, but it's good to see what our definition yields now. If n is odd, then r_n equals the proportion of zeros of the first $n + 1$ elements, which are exactly half, so $r_n = \frac{1}{2}$. If n is even, then among the first $n + 1$ elements of alt the number of zeros is just one more than the number of ones. So the sequence r_n is as follows:

$$1, \frac{1}{2}, \frac{2}{3}, \frac{1}{2}, \frac{3}{5}, \frac{1}{2}, \frac{4}{7}, \frac{1}{2}, \frac{5}{9}, \ldots$$

In short, for every n the number r_n is either equal to $\frac{1}{2}$, or slightly more, but the difference with $\frac{1}{2}$ is always less than $\frac{1}{n}$. For any $\epsilon > 0$ we choose a number n with $n > \frac{1}{\epsilon}$, then $\frac{1}{2} - \epsilon < \frac{1}{2} \leq r_n < \frac{1}{2} + \frac{1}{n} < \frac{1}{2} + \epsilon$, and this proves that $[alt]_0 = \frac{1}{2}$ according to the definition of limit.

Similarly, more generally one proves that for any periodic sequence u^∞ and any symbol x one has

$$[u^\infty]_x = \frac{|u|_x}{n},$$

where $n > 0$ is the length of u, and $|u|_x$ is the number of occurrences of x in u. Note that this is a *rational number*, since both $|u|_x$ and n are natural numbers. Combined with Theorem 5.3, this remains true if we add any random word to the front, thus obtaining for any ultimately periodic sequence:

Theorem 5.4 *For every ultimately periodic sequence a and every symbol x, the frequency $[a]_x$ exists and is a rational number.*

Theorems 5.3 and 5.4 explicitly state that for certain sequences a the frequency $[a]_x$ of the symbol x *exists*. This raises the following question: are there sequences a for which $[a]_x$ does not exist?

The answer turns out to be 'yes'!

As an example consider the following sequence a:

$$a = 0101001100001111000000011111111 \cdots$$

with alternating groups of zeros and ones, but the lengths of these groups become ever larger powers of 2: first twice a zero and a one, then two zeros and two ones, then four zeros and four ones, then eight, then 16, and so on. This sequence a may be defined more precisely by stating $a = 01b$ and $b = 01f(b)$, where f is the morphism defined by $f(0) = 00$ and $f(1) = 11$. Now let's look at the definition of $[a]_0$ Then we have to take the limit of the sequence $r_0, r_1, r_2, r_3, r_4, \ldots$ where

$$r_n = \frac{|w_{n+1}|_0}{n+1};$$

where $|w_{n+1}|_0$ is the number of zeros in the first $n+1$ elements of a.

If $n+1 = 2^k$, we see that the first $n+1$ elements are completely constructed by having a group of zeros followed by an equally large group of ones. So exactly half of the symbols is equal to 0, so $r_n = \frac{1}{2}$. This holds for every k.

Next, take $n+1 = 2^k + 2^{k-1} = 3 \times 2^{k-1}$. Then among the first 2^k elements exactly 2^{k-1} are equal to 0, extended by a group of 2^{k-1} elements that are all zeros. So we obtain

$$r_n = \frac{2^{k-1} + 2^{k-1}}{3 \times 2^{k-1}} = \frac{2}{3}.$$

And this applies to every k too. So we see that among the values $r_0, r_1, r_2, \ldots$ infinitely often the number $\frac{1}{2}$ occurs, but also infinitely often the number $\frac{2}{3}$. Then the definition of limit shows that there is no such limit L. So we conclude that for this particular sequence a the frequency $[a]_0$ *does not exist*.

We are close to the end of this chapter now. In this chapter we have seen a number of results that show a relationship between rational numbers and ultimately periodic sequences: a real number is rational exactly if its decimal notation is ultimately periodic, and the frequency of each symbol in an ultimately periodic sequence is always a rational number. Although being nice results, it has become a chapter with only very few pictures. We promise that this lack of pictures will be compensated in the following chapters. And as we do in all chapters, also this chapter is concluded by a challenge.

Challenge: the marble box

Challenge:

Start by a box containing 30 marbles: 10 red, 10 white and 10 blue. Apart from that a large amount of spare marbles is available, of all colors as much as is needed. Now you do a number of steps in which you always remove two marbles from the box and put one back in again. These two marbles are taken out at random, without looking, but putting them back is done according to the following rules:

- If you take two red ones, put a white one from the amount of spare marbles back in the box.
- If you take two white ones, put a red one from the amount of spare marbles back in the box.
- If you take two blue ones, put one of them back in the box.
- If you take a red and a white one, put a blue one from the amount of spare marbles back in the box.
- If one of the two marbles you picked is blue, put the other one back in the box.

In every step the number of marbles in the box is decreased by one, so you can do 29 of these steps, and after that there is exactly one marble left in the box.

What is the color of that remaining marble in the box?

6 Turtle figures

In this chapter, we present a way to make a picture out of an infinite sequence, in the hope that such a picture shows something about the structure of the infinite sequence. An old saying says that a picture says more than a thousand words, and this claim has been confirmed in several areas. In many branches of science, visualization of the topic is essential to make progress. To give an example, if you view a piece of plant or animal material by a microscope, you may recognize individual cells in it. To properly understand what is going on in these plants and animals, it is crucial to understand the structure and performance of these cells. It turns out that if you process the preparations with all kinds of materials in order to color them, much more information from the microscope images can be extracted. It has been quite a quest to find out which (color) substances are most suitable for this, all with the aim to visualize the building blocks of the living cell. We now want to do something similar with infinite sequences: how to visualize such a sequence in such a way that the resulting picture gives some information on properties of the sequence?

We will focus on a very simple way to visualize such a sequence in a picture: a *turtle figure*. Let's start with a bit of history. In 1948, Gray Walter manufactured his first robots, named *Elsie* and *Elmer*. The idea was that these could be controlled by commands like *forward* and *left*. The robots themselves and the resulting movements somewhat resembled turtles, which is why the robots were also called *turtles*. A long time later, from 1959, the PDP-1 computer was developed at MIT. There were no monitors back then, programming languages in development were Cobol, Fortran, Algol and Lisp, and the programs were entered with punched paper tapes. In 1967, a programming language *Logo* was developed by Feurzeig, Solomon, and Papert.

DOI: 10.1201/9781003466000-6

The main purpose was to give children a sense of programming by controlling a robot with commands like *forward* and *turn right*. The first implementation of this on the PDP-1 was written in Lisp and was called *Ghost*. The robot to be controlled was also called *turtle* here. When monitors were introduced later, Logo versions were also developed for this, in which the turtle was no longer a physical robot, but was simulated on a screen. And that is what is still called *turtle graphics* today. Inspired by this, the resulting figures are called *turtle figures*. Controlling such a turtle by commands like *forward* and *left* is a great way to teach children how to program, because you can immediately see what is happening in a picture. Many successful variants of Logo have been developed for this purpose. A more recent project like *Hedy* focuses on teaching everyone including children how to program, completely online without having to download or install anything. It will be no surprise that Hedy also has these turtle commands. Learning programming for children is both fun and useful. It should be recommended to offer that to all children, just like everyone learns writing and arithmetic.

We are now going to define what exactly we mean by a turtle figure. We consider the turtle just as a point in the plane having some walking direction. That turtle may perform actions. The first action is *forward*: for a number x by the command forward(x) the turtle moves a distance x in the direction indicated by his walking direction. Next, for an angle h, expressed in degrees, left(h) turns the turtle's walking direction h degrees to the left. Similarly, a command right is introduced, but that doesn't really bring anything new, because right(h) does the same as left($-h$) or left($360-h$). A figure in the plane consisting of line segments is called a *turtle figure* if these line segments describe the path that the turtle follows through a successive series commands forward(x) and left(h) for various values of x and h. To give an example, let us start at the point A and the turtle walking direction points to the right. Then by the commands

forward(2)
left(90)
forward(1)
left(90)
forward(1)
left(90)
forward(2)
left(90)
forward(2)

    left(90)
    forward(3)
    left(90)
    forward(1)
the following turtle figure is drawn:

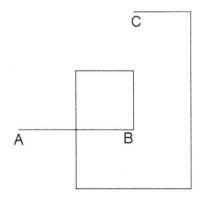

Indeed, we start in *A* and the turtle walking direction points to the right, so forward(2) draws a line segment of length 2 in that direction, ending up in point *B*. The next command is left(90), which turns the direction 90° to the left, making the walking direction upwards. Then by forward(1) a line segment of length 1 is drawn. And so on. At the end, with six times left(90), the walking direction points to the left, and by the last forward(1) the turtle figure arrives at point *C*. The resulting turtle figure consists of the seven line segments thus drawn with forward; the letters *A*, *B* and *C* are added for explanation only.

To get a reasonable size, here we have chosen the unit distance quite large. Choosing a unit ten times smaller, and multiplying all arguments of forward by ten yields exactly the same picture.

In this first example, all angles were straight, 90°, but other angles are allowed too, which will be exploited extensively later on.

Turtle figures of words and sequences

A word or sequence consists of a finite or infinite series of symbols. Such a word or sequence yields a turtle figure by linking one or more commands to each symbol, of the type left or forward, or perhaps even more complicated drawing commands, and then process them one by one. We want to keep it as simple as possible here. We also experimented with slightly more complicated shapes, for example, different distances per symbol, but that did not seem to substantially enrich the resulting turtle figures, so we focus on the simplest shape here.

For every symbol s, we choose an angle $h(s)$. Now the turtle figure of the word $u_0 u_1 \ldots u_{n-1}$ is defined to be the result of successively executing the commands left($h(u_i)$), forward(1), for $i = 0, 1, \ldots, n-1$. So for all symbols u_i in the word, first turn the walking direction $h(u_i)$ to the left and then draw a line segment of length 1 ahead.

The turtle figure just given can be obtained in this way by choosing $h(0) = 0$ and $h(1) = 90$, and the word 001110101001. Here we observe that we get the first forward(2) by combining the first two symbols 00 of the word, resulting in the four successive commands

$$\text{left(0)forward(1)left(0)forward(1)}.$$

This has exactly the same effect as forward(2) since left(0) only turns $0°$, so does not do anything. The next three ones in the word yield left(90)forward(1) three times. Next, the last forward(1) is extended to forward(2) by the subsequent 0 delivered by left(0)forward(1). Processing the remaining symbols of the word exactly completes the turtle figure as it was shown.

In the next example, the angles are no longer $0°$ or $90°$. Now we take the word 000000000, so nine zeros, and choose $h(0) = 160$. That means that the turtle figure is obtained by executing the two commands left(160) and forward(1) nine times, that is, turning $160°$ to the left each time, making a sharp angle of $20°$, and move forward one step of length 1. This produces the following star figure with nine points: after nine steps the turtle is exactly back where it started, having the same walking direction.

More general, this trick works for any rational angle: if $h(0) = \frac{360m}{n}$, and we take the word 0^n consisting of n zeros, then after n steps, the turtle is exactly at the same point where it started, having the same walking direction. The above star example is obtained by choosing $n = 9$ and $m = 4$, yielding $h(0) = \frac{360 \times 4}{9} = 160$.

In the next example take $h(0) = 141 = \frac{360 \times 47}{120}$. Then the turtle is back at the beginning by taking the word 0^{120}. The next picture shows the turtle figure for the word 0^{100}: well on the way, but not quite finished yet. As a puzzle you may try to find the beginning and the end of this turtle figure.

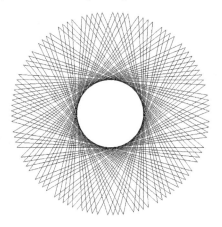

If we do complete it, so choosing $h(0) = 141$ and the word 0^{120}, then the turtle figure looks like this:

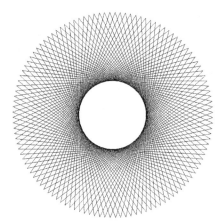

What happens if we choose the word 0^{130}, or 0^{1000}, or how many zeros we want, as long as there are at least 120? Then we keep exactly the same turtle figure, since after 120 steps the turtle is back where it started, both in position and direction, and every next zero will just draw a segment that has been drawn before. If this was done by a physical robot, small moving errors would be likely. But we do it on a computer, and after 120 steps it is exactly where it started, and the result shows no difference at all for a line segment drawn several times. Extending this forever, we observe that even for the infinite sequence 0^∞ its turtle figure coincides with the same finite picture. Really drawing the turtle figure of an infinite sequence would take infinitely long. But in this case we know that after a finite initial part of the sequence no new segments will be drawn, we may stop after this finite part, and consider the resulting turtle figure as the turtle figure of the infinite sequence.

Later we will see turtle figures of other sequences, but for which more complicated arguments show that after a finite part no new segments will be drawn, so such a finite part suffices to obtain the full turtle figure. For other turtle figures that we will see the drawing process goes on forever and would yield an infinite size picture. But even then the turtle figures of a large finite initial part of the sequence may give interesting shapes.

Turtle figures of periodic sequences

The turtle figures we saw until were for words and sequences only consisting of zeros. Now we will look at words and periodic sequences over more symbols. Let's take the word $u = 01001$ and choose $h(0) = 30$ and $h(1) = -114$. Starting at point A and the walking direction pointing to the right gives the following turtle figure:

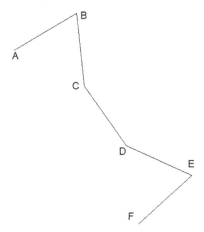

At the first 0 of 01001 the turtle starts in A, rotates 30° to the left and then does a step forward, ending in B. Then because of the first 1 in 01001 the direction turns 114° to the right and the turtle does another step forward, arriving in C. Then 30° to the left twice and one step forward which brings the turtle to D and E. Finally, another 114° to the right and one step forward, the turtle ends up in F. In total, the walking direction rotated 30° to the left three times and 114° to the right twice, that is a total of $3 \times 30 - 2 \times 114 = 90 - 228 = -138°$ to the left, or 138° to the right. Continuing this, the turtle figure of $u^2 = 0100101001$ is as follows, starting in the same way, but then from F followed by a rotated copy of the same picture:

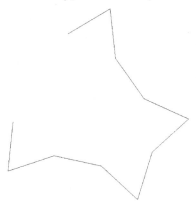

We observe that $138 = \frac{360 \times 23}{60}$, so for u^n the resulting direction is different from the starting direction for $n < 60$, but for u^{60} the direction is exactly equal to the starting direction. And not only that, by u^{60} the turtle ends up exactly at the starting point A again; later we will explain why. The turtle figure for u^{60} looks like this:

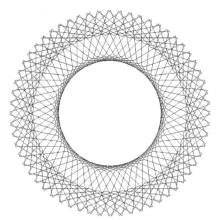

But this is not only the turtle figure for u^{60}, but also for u^n for every $n \geq 60$, because after these first 60 rounds every u draws the same five line segments that were drawn 60 rounds earlier. So this is not only the turtle figure for u^n for every $n \geq 60$ but also for the infinite sequence u^∞. We will prove later in Theorem 6.1 that this holds in general.

By varying the word u consisting of zeros and ones, and the angles $h(0)$ and $h(1)$, a lot of similar step figures are obtained. A second example is the turtle figure of u^∞ for $u = 01110011$ and $h(0) = 120$ and $h(1) = -170$:

The sharp angles on the outside are angles of $10°$, and are due to $h(1) = -170$; in the middle we see a kind of lace work like pattern.

A next example is the turtle figure of u^∞ for $u = 010001111$ and $h(0) = 120$ and $h(1) = -147$:

So far in all examples the angles $h(0)$ and $h(1)$ had an integer number of degrees, but they don't have to. In the following example we have $u = 0100101$,

$$h(0) = \frac{1845}{16} = 115\frac{5}{16} \quad \text{and} \quad h(1) = -\frac{2745}{16} = -171\frac{9}{16} :$$

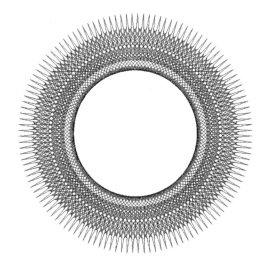

In all these examples, for u^n we are back at the beginning for some n, both in position and direction, and for larger n only existing line segments are redrawn without adding anything new. But why is that the case? And at what angles does all this happen? That's what we're going to find out now.

Some theory

To investigate some general observations about turtle figures, we first need some notations. This allow us to give a precise definition of turtle figure.

First, it is useful not only to choose $h(0)$ and $h(1)$, and maybe also $h(2)$ and $h(3)$ if the symbols 2 and 3 also participate but also define $h(u)$ for every word u. We do this as follows: if $u = u_0 u_1 \ldots u_{n-1}$, then

$$h(u) = h(u_0) + h(u_1) + \cdots + h(u_{n-1}),$$

that is, the sum of the angles of all symbols in u. For each u, the angle $h(u)$ indicates exactly the direction of the turtle after drawing the turtle figure of u.

That's about the angle. But what position does the turtle end up after drawing the turtle figure of a word u?

A position in the plane is determined by a pair (x, y), where x and y are two real numbers. First a special point is fixed, namely $(0,0)$, called the *origin*. In the case of turtle figures, we choose this origin to be the position of the starting point of the turtle. It is common to use x for the horizontal direction, and y for the vertical direction. This is just a choice, but there's no reason to deviate from this common usage, so we won't. This looks like:

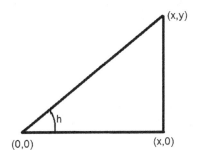

If we move a distance x to the right from $(0,0)$, we end up in $(x, 0)$. If we go up a distance y from $(x, 0)$, we end up in (x, y). In this picture x and y are both positive numbers, but they can also be negative, then x goes left and y goes down. In this way, every point in the plane can be written as (x, y) where x and y are real numbers. Such a point relative to the origin is called a *vector*. A nice feature is that now such vectors may be added, or multiplied by a number: we define

$$(x, y) + (x', y') = (x + x', y + y') \text{ and } c(x, y) = (cx, cy)$$

for all real numbers $x, y.x', y', c$. Because of $(x, y) + (-x, -y) = (0, 0)$ we also write $-(x, y) = (-x, -y)$.

More general, vectors may be considered in more than two dimensions, even in infinite dimensions. With our focus on turtle figures, we keep it simple and restrict to two dimensions.

In the picture we indicated the angle h between the horizontal line from $(0,0)$ to $(x, 0)$, and the line from $(0,0)$ to (x, y). There is a relationship between this angle h and the values x and y. If the length of the segment from $(0,0)$ to (x, y) is exactly 1, then y is equal to the *sine* of h, denoted by $\sin(h)$, while x equals the *cosine* of h, denoted by $\cos(h)$. These sine and cosine play an important role in the mathematics of the plane as most readers may have learned in school. If you haven't seen it before, or you've forgotten about it, don't worry, because all we will use is the feature just described. We just note that if the length of the segment from $(0,0)$ to (x,y) is not 1, but let's say ℓ, then everything is multiplied by ℓ, and we have $x = \ell\cos(h)$ and $y = \ell\sin(h)$.

Let's go back to turtle figures. Just as we denote $h(u)$ for the angle after processing the word u, we denote $P(u)$ for the position where we end up after processing this word. Let's see what that looks like for a word $u = u_0u_1u_2$. Starting with angle 0, which is the direction pointing horizontally to the right, by processing u_0, the angle $h(u_0)$ is added to 0, followed by a unit step in that direction. Then according to the picture above, the turtle arrives at (x,y) where $x = \cos(h(u_0))$ and $y = \sin(h(u_0))$. So:

$$P(u_0) = (\cos(h(u_0)), \sin(h(u_0))).$$

Then we process u_1 from that new point: $h(u_1)$ is added to the angle, followed by a unit step in the new direction, ending up in

$$P(u_0u_1) = (\cos(h(u_0)) + \cos(h(u_0) + h(u_1)), \sin(h(u_0)) + \sin(h(u_0) + h(u_1))).$$

Finally, we process the third and final symbol u_2 in the same way and arrive in

$$P(u) = P(u_0u_1u_2) = (x, y),$$

where

$$x = \cos(h(u_0)) + \cos(h(u_0) + h(u_1)) + \cos(h(u_0) + h(u_1) + h(u_2))$$

and

$$y = \sin(h(u_0)) + \sin(h(u_0) + h(u_1)) + \sin(h(u_0) + h(u_1) + h(u_2)).$$

For longer words, it works the same way: for each subsequent symbol s the angle h_s is added to the direction, adding a cosine to the x value and a sine to the y value, more precisely

$$h(us) = h(u) + h(s)$$

and

$$P(us) = P(u) + (\cos(h(u) + h(s)), \sin(h(u) + h(s))),$$

for every word u and every symbol s. To understand what is happening here, it may help to use the term *rotation* over a certain angle. To find $P(us)$ for a word u and a symbol s, we first find the vector $P(u)$, to which the vector $(\cos(h(u) + h(s)), \sin(h(u) + h(s)))$ is added. This added vector is a rotated version of $P(s) = (\cos(h(s)), \sin(h(s))))$, rotated over the angle $h(u)$. Denote rotation over an angle h by R_h, then we obtain

$$P(us) = P(u) + R_{h(u)}(P(s)).$$

This applies more generally if we replace the symbol s by any word v: to determine $P(uv)$ we first determine $P(u)$ and then add the rotated version of $P(v)$. We will use this property more often later, so let us present it by a lemma:

Lemma 6.1 *For all words u, v with corresponding angles we have*

$$P(uv) = P(u) + R_{h(u)}(P(v)).$$

Now we are to define exactly the turtle figure of a word or sequence, if we have chosen an angle $h(s)$ for each symbol s. The *point sequence* of the turtle figure is the sequence

$$P_0, P_1, P_2, \ldots$$

of points where $P_i = P(u_i)$ for each i, and u_i is the word consisting of the first i symbols of the word or sequence. For a word of length n, this sequence consists of $n+1$ points because it is defined up to and including $i = n$. For an infinite sequence, the point sequence is also infinite. Now the *turtle figure* is the figure obtained by connecting every two consecutive points in this point sequence by a line segment.

Finite turtle figures of periodic sequences

If $h(u)$ is rational and not zero, then the turtle figure of the periodic sequence u^∞ consists of only finitely many line segments. In this section, we will formulate this in a theorem and prove it. Such a turtle figure consisting of finitely many line segments is called a *finite turtle figure*. Before we give and prove the theorem, we first collect some general properties of rotations in a lemma, which is just an auxiliary theorem:

Lemma 6.2 *Let (x, y) and (x', y') be two points, and h and h' be two angles. Then we have*

1. $R_{h+h'}(x, y) = R_h(R_{h'}(x, y))$,
2. $R_h(x + x', y + y') = R_h(x, y) + R_h(x', y')$,
3. *If $n \times h = 360k$ for natural numbers n, k then $R_{nh}(x, y) = (x, y)$.*

All these properties are easily proved from the definitions, we do not go into further detail.

Rational numbers are by definition of the form $\frac{m}{n}$ where m is an integer and n is a natural number > 0. If we look at rational angles, we may assume that $m \geq 0$, otherwise we may add $360°$ just as many times until it is ≥ 0. Since $360°$ describes one full turn, it is convenient to write an arbitrary rational angle as $\frac{360m}{n}$. Remember that an angle remains the same if we add or subtract $360°$. In the following, we will therefore often identify an angle that is a multiple of $360°$ by 0.

Theorem 6.1 *Let u be a non-empty word of length $k > 0$, and let the angle $h(s)$ be defined for every symbol s occuring in u, by which also $h(u)$ is defined. If $h(u) = \frac{360m}{n}$ and $h(u) \neq 0$, then the turtle figure of u^∞ is equal to the turtle figure of u^n, hence finite, and consists of at most kn distinct line segments.*

Proof: Since k is the length of u, u^n has length kn, and the turtle figure of u^n consists of kn line segments, some of which may coincide, so at most kn of them are distinct. We now show that the turtle figure of u^∞ consists of exactly the same line segments, so that after processing u^n only line segments will be drawn that we already had before. Later we will show that $h(u^n) = 0$ and $P(u^n) = (0,0)$. Using this, it follows that $P(u^n v) = P(v)$ for each word v, and it follows that the next series of kn line segments in the turtle figure of u^∞ correspond to the first series of kn line segments. This argument is repeated: for every i we obtain $h(u^{in}) = 0$ and $P(u^{in}) = (0,0)$, and every subsequent series of kn segments corresponds to the first set of kn segments. This proves the theorem, if we moreover show that $h(u^n) = 0$ and $P(u^n) = (0,0)$.

That the statement $h(u^n) = 0$ holds, follows from $h(u^n) = nh(u) = 360m$, and as a multiple of $360°$ that is equal to 0.

It remains to show that $P(u^n) = (0,0)$. We do this by showing that $R_{h(u)}(P(u^n)) = P(u^n)$. Because $h(u) \neq 0$, the rotation $R_{h(u)}$ rotates over an angle $\neq 0$, for which $(0,0)$ is the only point that rotates towards itself.

We have already seen in Lemma 6.1 that $P(uv) = P(u) + R_{h(u)}(P(v))$ for all words u, v. If we apply this repeatedly we get

$$P(u^n) = P(u) + R_{h(u)}(P(u)) + R_{2h(u)}(P(u)) + \cdots + R_{(n-1)h(u)}(P(u)).$$

Now we will apply $R_{h(u)}$ to this. Using Lemma 6.2, part 2, we then get

$$R_{h(u)}P(u^n) = R_{h(u)}(P(u)) + R_{h(u)}(R_{h(u)}(P(u)))) +$$

$$R_{h(u)}(R_{2h(u)}(P(u))) + \cdots + R_{h(u)}(R_{(n-1)h(u)}(P(u))).$$

Using Lemma 6.2, part 1, we then get

$$R_{h(u)}P(u^n) = R_{h(u)}(P(u)) + R_{2h(u)}(P(u)) + R_{3h(u)}(P(u)) + \cdots + R_{nh(u)}(P(u)).$$

Now notice that $n(h(u) = 360m$, so according to Lemma 6.2, part 3, the last term $R_{nh(u)}(P(u))$ in this sum is equal to $P(u)$. So these n terms are exactly the same as in

$$P(u^n) = P(u) + R_{h(u)}(P(u)) + R_{2h(u)}(P(u)) + \cdots + R_{(n-1)h(u)}(P(u)),$$

only all shifted one position, but yielding the same sum. So $R_{h(u)}(P(u^n)) = P(u^n)$. Since $(0,0)$ is the only point that rotates towards itself, we have $P(u^n) = (0,0)$, completing the proof of the theorem. $\square$

Infinite turtle figures of periodic sequences

In Theorem 6.1, the main conclusion is that the turtle figure is finite. Two crucial conditions were actually used in the proof: $h(u)$ is rational and $h(u)$ is non-zero. In mathematics, in such a case, it is a good principle to investigate whether such conditions are really essential: if the conditions do not hold, can we find a counterexample, or would the stronger theorem hold without requiring the conditions? In this case, we will show that both conditions are essential: if they do not hold, then the statement is not true, and we may get infinite turtle figures.

Let's first consider the requirement $h(u) \neq 0$. The simplest case we can think of is immediately successful: let's choose $u = 0$ and $h(0) = 0$. Then the turtle figure of $u^\infty = 0^\infty$ consists of an infinite straight line, and that consists of infinitely many line segments of which no two coincide. A picture of this is not so exciting; it is more interesting to give a turtle figure of a slightly less trivial example with two symbols and angles that are not 0. Here we choose $u = 0101010111$ and $h(0) = 45$ and $h(1) = 90$. Since u consists of four zeros and six ones, we get $h(u) = 4 \times 45 + 6 \times 90 = 720 = 2 \times 360$, indeed corresponding to $0°$. The turtle figure of u looks like this:

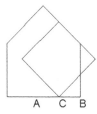

The starting point here is point A, and the starting direction is chosen so that the first line segment, the turtle figure of 0, is the horizontal line segment AB. After the ten steps of $u = 0101010111$ the turtle has reached point C. For the turtle figure of u^∞, this picture is drawn infinitely many times, shifting each subsequent copy a distance of AC to the right. This yields the infinite turtle figure that begins as follows, and continues infinitely to the right:

This allows an endless amount of variations: using the same $u = 0101010111$, but as angles $h(0) = 135$ and $h(1) = 150$ the next figure is obtained, continuing in a similar fashion infinitely far to the right:

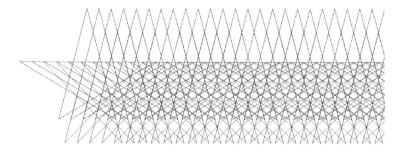

These examples show that in Theorem 6.1 the requirement $h(u) \neq 0$ is essential, since $h(u) = 0$ in these examples and the conclusion of Theorem 6.1 does not hold. But what about the other condition on rationality of $h(u)$? If $h(u)$ is not rational, the conclusion of Theorem 6.1 does not hold either, but for a completely different reason. The turtle figure of u^∞ then consists of infinitely many different line segments, but they all remain close to each other. So they do not go infinitely far away as in the examples we saw with $h(u) = 0$. Let's give an example: choose $u = 0$ and $h(u) = 180\sqrt{2}$. Then the turtle figure of 0^{100} looks as follows:

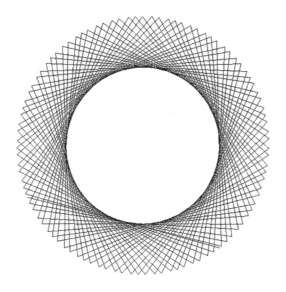

Here we see 100 line segments. Careful inspection may show the beginning and the end of the turtle figure. But if we go beyond 100 we never get
back to the starting point for the following reason. If after n steps the turtle
is back at the starting point with the walking direction coinciding with the
initial direction, then $n \times h(0)$ is a multiple of $360°$. But then $h(0)$ is rational,
and $h(0)$ was chosen to be not rational. Conclusion: the turtle figure of 0^∞
continues to rotate in circles indefinitely in the above picture, but draws a
new line segment in every step. Not all points are reached, but a set of points
that is *closed* in the drawn area, that is, for every $\epsilon > 0$ every point of the
set is reached at distance $< \epsilon$. If for ϵ we choose the size of a pixel, it means
that all the pixels in that area become black. To save ink, we show the turtle
figure of the full sequence 0^∞ slightly smaller, being just a solid black band:

If we choose another u with $h(u)$ not rational, we get a similar turtle figure
of u^∞ consisting of infinitely many distinct line segments.

Ultimately periodic sequences

The turtle figure of a ultimately periodic sequence vu^∞ is as follows: the figure starts by the turtle figure of the word v, and then, starting at the end point and the angle obtained, is continued by the turtle figure of u^∞. With the examples in mind we have already seen, it is easy to obtain a turtle figure of u^∞ looking like a symmetrical flower figure. For example, we get this for $u = 0102$, and $h(0) = 24$, $h(1) = 156$, $h(2) = 126$. Here $u = 0102$ draws a petal, and u^∞ yields a flower of 12 such petals, because $12h(u) = 12(2 \times 24 + 156 + 126) = 12 \times 330$ is a multiple of $360°$. Using these angles, the word $v = 00010200102001020000$ yields a turtle figure consisting of a stem with a few leaves attached. Combining these two ingredients yields the turtle figure of the ultimately periodic sequence $vu^\infty = 00010200102001020000(0102)^\infty$:

The turtle figure of an ultimately periodic sequence vu^∞ consists of finitely many line segments exactly as it holds for u^∞, because v itself always gives finitely many line segments, namely at most j if j is the length of v. Combined with Theorem 6.1 this yields the following theorem:

Theorem 6.2 *Let u be a non-empty word of length $k > 0$ and let v be a word of length j. Let the angle $h(s)$ be defined for every symbol s that appears in*

u or v. If $h(u) = \frac{360m}{n}$ and $h(u) \neq 0$, then the turtle figure of vu^∞ equals the finite turtle figure of vu^n, and consists of at most $j + kn$ distinct line segments.

And indeed, in the flower figure just shown, v has length $j = 20$, and $k = 4$ and $n = 12$, and the number of line segments in the turtle figure is 67, just count. That is one less than the bound given in Theorem 6.2, since the last segment produced by v coincides with a segment from u^∞. This confirms the *at most* statement in the theorem for this example.

Now we will give the solution of the marble box challenge from Chapter 5. A first guess may be that the color of the last marble depends on the way the two marbles are picked form the box in all of the rounds. Surprisingly, that is not the case. Let's see what happens to the number of marbles in the box of each of the three colors at each of the given five types of steps. The following table shows what happens to the numbers of red, r, white, w, and blue, b, marbles in the box. Here, 0 indicates that the number remains the same, $+1$ that one is added, -2 that two are subtracted, and so on.

kind of step	r	w	b	$r - w$
1	-2	$+1$	0	-3
2	1	-2	0	$+3$
3	0	0	-1	0
4	-1	-1	$+1$	0
5	0	0	-1	0

An extra column has been added to this table: what happens to $r - w$, that is, the number of red minus the number of white marbles in the box? For example, in the first case the number of red marbles decreases by two and the number of white marbles increases by one, so $r - w$ then decreases by three.

Here we do a remarkable observation: in every step the number $r - w$ either increases or decreases by three, or stays the same. Initially we have $r = w = 10$, so $r - w = 0$. Hence the number $r - w$ always remains a multiple of 3, positive or negative, at any number of steps. That gives the solution to the challenge: after 29 steps $r - w$ is a multiple of 3. Since there is only one marble left in the box, this should be blue, since otherwise it is red or white for which $r - w = 1$ or $r - w = -1$, none of which is a multiple of 3. So the answer of the challenge is: the last marble in the box after 29 steps is blue.

This type of reasoning uses an *invariant*, that is, a property that holds at the beginning, holds after every step, and therefore holds at the end after any possible number of steps. In this solution the invariant is the property that $r - w$ is a multiple of 3.

Invariants are widely used to show that some property does not occur, namely reaching a situation where the invariant does not hold. If you suspect that the answer to the paint pot problem is 'no', you may try to prove it using an invariant.

This is also a commonly used technique for proving that a program is *correct*. Then the key is to find an invariant of the program that shows that unwanted behavior does not occur.

Challenge: subword with zero angle

We say that a word v is a *subword* of a word u if the word u can be written as $u = u'vu''$ for words u' and u'', that are allowed to be empty. So a subword of u is obtained by removing zero or more symbols from both the beginning and the end of u.

> **Challenge:**
>
> Let u be a word of length 360 over an alphabet of 7 symbols. Let for each of the 7 symbols s the angle $h(s)$ be an integer, measured in degrees. Establish whether the following is always true: the word u has a non-empty subword v such that $h(v) = 0$, up to multiples of 360.

To give a hint: for a word of length 359 the claim is not true. In that case we may choose $u = 0^{359}$ and $h(0) = 1$. Then every subword of u only consists of zeros, so $h(v)$ is equal to the length of v. For a non-empty subword v of u then the angle $h(v)$ is at least 1 and at most 359, so not zero. However, this particular example no longer works for length 360, because then we may choose the subword v to be equal to u of length 360, for which we have $h(v) = 360 = 0$.

 # Programming

We have already seen quite some pictures of turtle figures, and many more will follow. All these turtle figures are created by a computer program. In this chapter, we will describe how this works.

Nowadays, a considerable part of the world's population is used to using computers. If you want to enter text and create a document, you use a word processing program, and if you want to look up something on the Internet, you use a browser and a search engine. But that's all about using existing programs. Creating your own programs, *programming*, is much less common. That's a pity, because programming is so much fun, and not difficult at all. I am sure that also for non-professional programmers the importance goes beyond just being fun: programming changes your attitude towards using computers: by programming you are much more in charge of what your computer does.

The core of programming is that you write down commands and instructions on how to execute them, and then the computer actually executes them. To write down these commands and instructions, you need a *programming language*, usually within a *programming environment*. Several programming languages and programming environments are available. They have a lot in common: many basic constructions appear in many programming languages, often with small variations in notation. A typical command might be to give a value to a variable, and a typical statement would be a *for* that allows you to execute such a command or series of commands a large number of times.

When drawing pictures in *turtle graphics*, basic commands may include left and right to turn the turtle a certain angle to the left or right. In some

DOI: 10.1201/9781003466000-7

programming languages, the command may be `turn`, and by using negative numbers as well, this is used for turning both left and right. Another basic command is `forward` to move the turtle a certain distance in its direction. These basic commands are sufficient for the turtle figures that we encounter in this book.

Turtle programming in Python

Several programming languages offer this *turtle graphics* directly. One of them is *Python*, a very accessible programming language, used all over the world. It has been developed by the Dutchman Guido van Rossum. This language and its environment are freely available and easy to install on any modern computer; how to do so is easily found on the internet. In Python, our turtle figures are easy to program and draw. As an example, we show how to draw the turtle figure of $(011)^\infty$ for $h(0) = 170$ and $h(1) = 60$. According to Theorem 6.1 and the fact that $36 \times h(011) = 36 \times 290 = 29 \times 360$, we know that the turtle figure of $(011)^\infty$ is equal to the turtle figure of $(011)^{36}$. So the program has to draw the turtle figure of 011 for 36 times. The basic Python command

```
for i in range(36):
```

is used to execute what follows 36 times. More precisely: for the variable *i* ranging from 0 to 35. For our turtle figure, we want the part that is executed 36 times to consist of three times `left`, always followed by `forward`. Each `left` and `forward` still has to be given a number to indicate how much to turn to the left, and what distance to move forward. Such a value that is passed is called a *parameter*. A fixed number may be used for this, for example `left(60)`, but if the same parameter is used more often, or changes later, it is useful to enter a *variable* instead. In our case, we introduce variables `d`, `h0`, and `h1` for the values of the distance and the angles $h(0)$ and $h(1)$. The values $h(0) = 170$ and $h(1) = 60$ were already given. For choosing a suitable distance, some experimenting is required to find a value producing a picture that is not too big and not too small. Here it turns out that $d = 180$ gives a nice result. To make the *turtle graphics* available, we need to include `from turtle import *` at the beginning. By the command `done()` the drawn picture is not lost at the end.

The full Python program now looks like this:

```
from turtle import *
d = 180
h0 = 170
h1 = 60
for i in range(36):
    left(h0)
    forward(d)
    left(h1)
    forward(d)
    left(h1)
    forward(d)
done()
```

When running this program, slowly the following turtle figure will be drawn:

On the right side of this picture, we see a triangle, being the turtle in the final position, which is equal to the starting position.

It is not the purpose of this book to include a full Python programming course. We only want to give an impression of what such a program looks like. The main message is that this program directly reflects the instructions as they were given in words. A note about indentation makes sense: the whole part from `left(h0)` to the last `forward(d)` is indented, and this means that this is the part that is executed 36 times by the command

`for i in range(36):`

Other programming languages may use other notations. For example, this grouping together is denoted by enclosing the group in braces in languages such as C and Java. In Pascal, the same feature is denoted by enclosing the group in `begin` and `end`.

Let's give another example of a Python program that draws the turtle figure of a periodic sequence. Now choose the sequence $(100101110)^\infty$ and angles $h(0) = 0$ and $h(1) = 151$. According to Theorem 6.1 the turtle figure of the infinite sequence is equal to that of $(100101110)^{72}$. Doing it in the same way as above would yield nine times `left` and nine times `forward` in the group inside the `for` command. But for every $h(0) = 0$, we may omit the `left` since turning over an angle 0 doesn't do anything. And a number of consecutive `forward` commands may be replaced by a single `forward` command with a larger parameter. Doing so yields the following program:

```
from turtle import *
d = 160
h = 151
speed(0)
for i in range(72):
        left(h)
        forward(3*d)
        left(h)
        forward(2*d)
        left(h)
        forward(d)
        left(h)
        forward(d)
        left(h)
        forward(2*d)
done()
```

Moreover, we added the command `speed(0)` in order to do the drawing at maximum speed, which is not really very fast yet. The result is the following, in which the small triangle on the right again indicates the turtle at the end:

Turtle programming in Lazarus

Python is not the only language having features for turtle programming. In fact, almost all turtle figures in this book were created using the programming environment *Lazarus*. One reason not to use Python anymore is that we want to be able to draw much larger turtle figures, up to hundreds of thousands of line segments, within a few seconds. This works fine in Lazarus. Sure it would work fine in several other environments too, but we chose Lazarus.

Lazarus is an *open source* variant of the programming environment *Delphi*. Here *open source software* means that it is freely available, including the source code. In this way in principle, even variants and extensions of the programming environment may be made. This environment Delphi is an extension of the language *Pascal*. This was the language in which I learned programming quite some time ago, in 1976. At that time, this was running on a large central computer at the university. Programming was done by first typing your program on a stack of punch cards. This stack of punch cards was handed in at the desk of the computer centre. Next, after a few hours, the program was executed and the paper output could be collected. Typically, some minor typing error was made, resulting in output only consisting of an error message. This made you check all your punch cards very carefully before handing them in. So the handling was very different from the current way of programming, but the thrill that the computer carefully executed all millions of steps that I told him to do by my program, without any complaint, was exactly the same as a first programmming experience nowadays.

A main feature in the expansion from Pascal to Lazarus is in the graphics. Where a Pascal program is just a series of commands to be executed, in Delphi or Lazarus you can make programs with a *graphical user interface* (a *GUI*), in which you can click virtual buttons and have many other graphical facilities at your disposal. All kinds of drawing commands are available. These drawing commands are very useful to draw turtle figures. Two basic drawing commands are moveto and lineto, both of which always take two numbers as parameters. If you want to draw a segment from point $(x1, y1)$ to point $(x2, y2)$ for certain values of $x1, y1, x2, y2$, this is done by:

```
moveto(x1,y1);
lineto(x2,y2);
```

If you then execute lineto(x3,y3), without another moveto in between, then a line segment is drawn from the last point, i.e. $(x2, y2)$, to the new point $(x3, y3)$. This may be repeated as many times as you like: exactly what is needed when drawing turtle figures. If the point sequence of a turtle figure is

$$(x0, y1), (x1, y1), (x2, y2), (x3, y3), \ldots$$

then the corresponding turtle figure is drawn by the following sequence of commands:

```
moveto(x0,y0);
lineto(x1,y1);
lineto(x2,y2);
lineto(x3,y3);
...
```

How do we determine the positions $(x0, y0), (x1, y1), (x2, y2), (x3, y3), \ldots$ without the commands left and forward? This may be done by using sine and cosine, as discussed earlier. First, we enter a variable h that represents the angle of the turtle. If we want to rotate the turtle through an angle $h0$, denoted earlier by left(h0), this is now done by the command

```
h := h + h0;
```

This is Pascal's (and Delphi's and Lazarus') notation for giving the value $h + h0$ to the variable h. In Python (and also in C and Java) instead of := only = is denoted.

This allows us to deal with left. But how about forward? We need sine and cosine to determine the change in position. Let's use variables x and y where (x, y) represents the position of the turtle. So what should happen when executing `forward(d)`? Then we need to determine the turtle's new position, and draw a line segment to it. The following figure shows what that new position looks like:

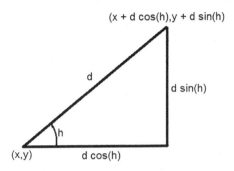

We assume that the command `lineto(x,y)` was performed in the previous step, by which a `moveto` command is not necessary. So we get the desired behavior of `forward(d)` by running the following three commands in sequence:

```
x := x + d *cos(h);
y := y + d *sin(h);
lineto(x,y);
```

To get everything working, one minor adjustment is needed. The result of sine and cosine is a *real* number, so x and y would also be real numbers. But for `lineto` the arguments must be integers, since they refer to the pixels as they appear on the screen. This is easily remedied by the operation `round` which rounds real numbers. So the last line `lineto(x,y);` has to be replaced by

```
lineto(round(x),round(y));
```

and then it works perfectly. This is how all turtle figures in this book have been created, except for the few Python examples.

A full Lazarus program to create turtle figures of morphic sequences, including all sources, is freely available at

```
https://github.com/hzantema/turtle-graphics
```

By this program you may create all turtle figures of morphic sequences that are given in this book, and any variation of them that you like.

Some theory

The title of this section suggests that it will be about theory, and in this chapter, one may expect theory on programming. But that is only partly true: there will be theory, but not about programming. Instead the theory will lead to the solution of the paint pot problem, being the first challenge. The question was whether it was possible to start by a word of the shape *bcdeu* where *u* is an unknown and perhaps very long word, and convert that to a word that starts in *a*, so of the shape *av* for some word *v*. In this converting two kinds of steps are allowed: if *p* and *q* are consecutive distinct letters *b, c, d, e*, they may be swapped. So the word *pq* may be replaced by the word *qp*. This is the first kind of step. The second kind of step is: if *p* is one of the letters *b, c, d, e*, and both its neighbors are *a*, this word *apa* may be replaced by *pap*. This step may be reversed: *pap* may be replaced by *apa*. We already observed that in a simplified form it was possible: we showed how to convert some word starting in *bcd* to a word that starts in *a*. This was not very simple: the word was 12 symbols long, and the conversion took 15 steps.

Now we will show that for the real paint pot problem, so starting in *bcde*, it is not possible. For doing so, we will apply a completely different technique than finding a solution. The key lies in the following magic *model M* consisting of eight elements that we number from 1 to 8.

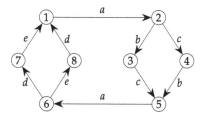

Here, the word *model* just means a collection of elements, eight in this case. In this model, we are going to interpret each word over the letters *a, b, c, d, e* as one of these eight elements. Then our main result will follow from the observation that the interpretations of words of the shapes *bcdeu* and *av* are always distinct.

For defining this interpretation, we first define a function $p_M : M \rightarrow M$ for each of the five letters *p* being *a*, *b*, *c*, *d* or *e*. If the element *i* in *M* has an arrow in the above picture labeled by *p* to the element *j* of *M*, then $p_M(i) = j$. If *i* has no such outgoing arrow, then $p_M(i) = i$. For example, there is an arrow labeled by *a* from 1 to 2, so $a_M(1) = 2$, and likewise $a_M(5) = 6$, and for all other i, so for $i = 2, 3, 4, 6, 7, 8$ we have $a_M(i) = i$. In the same way b_M is defined: $b_M(2) = 3$, and so on.

We do this not only for single letters but also for words: if $u = u_0 \ldots u_{n-1}$, then $u_M : M \to M$ is defined by

$$u_M(i) = u_{0,M}(u_{1,M}(\ldots u_{n-1,M}(i) \ldots)).$$

For example, we have

$$(abc)_M(1) = a_M(b_M(c_M(1))) = a_M(b_M(1)) = a_M(1) = 2,$$

and

$$(abc)_M(2) = a_M(b_M(c_M(2))) = a_M(b_M(4)) = a_M(5) = 6.$$

Note that to determine $u_M(i)$ you start in i, then interpret the *last* element of u. Then you apply the interpretation of the penultimate element of u to that result, and this continues until you have processed all elements of u from last to first.

We could have defined such a model in many ways, with a different set of elements, and different labeled arrows. But this particular model has a very special property: if we take a step from the paint pot game from a word u to a word v, then $u_M(i) = v_M(i)$ for every i in the model. This is checked this for all i and for all types of steps. Checking all cases we leave to the reader, here we restrict to a few typical cases. For example, for steps of the first type where we swap consecutive letters b, c, d, e:

$$d_M(e_M(6)) = 1 = e_M(d_M(6)),$$
$$b_M(e_M(6)) = 8 = e_M(b_M(6)),$$
$$b_M(c_M(7)) = 7 = c_M(b_M(7)).$$

For steps of the second type, we have

$$a_M(b_M(a_M(1))) = 3 = b_M(a_M(b_M(1))),$$
$$a_M(d_M(a_M(8))) = 2 = d_M(a_M(d_M(8))),$$
$$a_M(c_M(a_M(6))) = 6 = c_M(a_M(c_M(6))).$$

This can be checked for all other allowed steps and all elements in the model.

This does not only apply to words of length 2 or 3, but to any word on which a step applies. For instance, if u and v are arbitrary words, then for the conversion step from $uabav$ to $ubabv$ we have

$$(uabav)_M(i) = u_M((aba)_M(v_M(i))) = u_M((bab)_M(v_M(i))) = (ubabv)_M(i),$$

and similar for every other allowed step. When taking any allowed step, the interpretation of the word therefore remains the same, and therefore also for several allowed steps in succession, no matter how many there are. Now we are ready to give the solution of the paint pot problem. Assume that there is a series of allowed steps from *bcdeu* to *av* for certain words *u* and *v*. Then, according to the above observation, we have

$$(bcde)_M(u_M(1)) = (bcdeu)_M(1) = (av)_M(1) = a_M(v_M(1)).$$

This will give rise to a contradiction. We don't know exactly which element $u_M(1)$ is, but it is one of eight elements 1–8. For each of these eight elements *i* we check that $(bcde)_M(i)$ is equal to either 1 or 5. So $(bcde)_M(u_M(1))$ is equal to 1 or 5.

On the other hand, the element $a_M(v_M(1))$ is of the form $a_M(i)$ for some *i*, and for that we check that for every *i* this is one of the elements 2, 3, 4, 6, 7 or 8, but not 1 or 5. As a consequence, $(bcde)_M(u_M(1))$ and $a_M(v_M(1))$ are not equal, while they would be equal since there is a series of allowed steps from *bcdeu* to *av*. So this is a contradiction. Hence, we conclude that it is *not* possible to convert a word starting in *bcde* with allowed steps to a word starting in *a*, which completely solves the paint pot problem.

When launching this problem in Chapter 1, it was already stated that it was very difficult. If there are readers who found a solution themselves, I am very interested, and I would like to get in touch with them. Once you see our final solution, it takes some effort to check that it is all correct, but this doable with some patience. But how do you find such a solution?

Just as we saw in the marble box challenge earlier, one way to show that you can't get from an initial situation to a final situation by only using a particular set of rules, is to find an *invariant*. This is a property that remains to hold when applying steps according to the rules of the game. If the invariant applies in the initial situation, and that invariant does not apply in the desired final situation, the conclusion is that this final situation cannot be reached when only applying the rules. In the area of my research this is a quite common approach, so when I was playing around with this problem, and got the feeling it wouldn't be possible, an obvious approach was to try some invariance argument.

Finding a model in which certain equalities apply occurs more often in related areas. My idea then was to find a finite model by which the required properties are checked for the finitely many elements. But how to find such a finite model? I admit that this was not at all a result of being very clever:

I just wrote a big formula that, for a fixed number n, gave the combination of all the requirements for a model with n elements. To determine whether such a formula has a solution, some programs called *SMT solvers* are freely available that do this, combining extensive computing power and a lot of built-in cleverness. By writing a program generating the corresponding formula for $n = 2, 3, 4, \ldots$, and applying the SMT solver Z3 to the result, this showed that up to seven there was no solution. But for $n = 8$ Z3 did find a solution, and this solution was interpreted as the solution just given.

The argument can be seen as an invariance argument: we start by a word of the shape $bcdeu$ where $(bcdeu)_M(1)$ equals 1 or 5, Next we show that this holds at every step, so it still holds at the end. So av cannot be reached, since $(av)_M(1)$ is neither equal to 1 nor to 5. The invariant in this reasoning is the property that the word applied to 1 in the given model always yields the element 1 or 5.

As this section is on theory, we now give one more solution of a theoretical challenge: the challenge on monotone functions. We will show that if $f, g : \mathbb{N} \to \mathbb{N}$ are both monotone, then always some n in $\mathbb{N}$ exists such that $g(f(f(n))) \geq f(g(n))$. This quite tricky question came up in my research in the early 1990s in the analysis of terminating computations. Later, this problem was also submitted as a candidate for a problem in the International Mathematical Olympiad. Eventually it was not selected for the intended Olympiad itself, but for the *shortlist*, being some standard list of exercise problems for the Mathematical Olympiad.

Assume that the statement is not true. Then there are monotone functions f, g for which $f(g(n)) > g(f(f(n)))$ for all natural numbers n. Our goal is to derive a contradiction. First note that for the monotone map g the property $g(n) \geq n$ holds for every n, this is easy to prove by induction to n: $g(0) \geq 0$ certainly holds, and if $g(n) \geq n$ then $g(n+1) > g(n) \geq n$, so $g(n+1) \geq n+1$.

Now we prove by induction on k that $g(n) \geq f^k(n)$ holds for every k, n. For $k = 0$ this holds since $g(n) \geq n$ as we just proved. Now we assume as the induction hypothesis that $g(n) \geq f^k(n)$ for every n and for a given k, and we have to prove that $g(n) \geq f^{k+1}(n)$, for every n. Due to the assumption $f(g(n)) > g(f(f(n)))$, for every n, and the induction hypothesis, we have

$$f(g(n)) > g(f(f(n))) \geq f^k(f(f(n))) = f^{k+2}(n) = f(f^{k+1}(n)).$$

Since f is monotonic, it follows that $g(n) \geq f^{k+1}(n)$, which is exactly what we had to prove to conclude that $g(n) \geq f^k(n)$ for every k, n.

Since f is monotonic, $f(n) \geq n$ holds for every n. It is not the case that $f(n) = n$ holds for every n since this would contradict the assumption $f(g(n)) > g(f(f(n)))$ for every n. So there is a number n such that $f(n) > n$. But then due to monotonicity of f we have an increasing infinite sequence of numbers

$$n < f(n) < f(f(n)) < f^3(n) < f^4(n) < f^5(n) < \cdots .$$

We just proved that all numbers in this sequence are $\leq g(n)$. So we have infinitely many distinct numbers all $\leq g(n)$: that gives the contradiction. This concludes the proof and gives the solution of this challenge.

The next two sentences should be skipped if you are not busy with the challenge of the last chapter. If A is even, then 1 is subtracted from B, otherwise everything remains unchanged. Then the process continues just before the challenge on googolplex.

Challenge: knight moves

Consider the following square of size 5×5:

25	8	3	14	23
10	15	24	19	4
7	2	9	22	13
16	11	20	5	18
1	6	17	12	21

The numbers from 1 to 25 appear in this square in a special way: for every number the move to its successor is a *knight move* known from chess: two steps horizontally and one step vertically or vice versa. Indeed: from 1 to 2 is a knight move, from 2 to 3 too, also from 3 to 4, and so on until the last step from 24 to 25. All these steps are knight moves, and every of the 25 cells is visited exactly once. Such a path consisting of successive knight moves never passing the same cell more than once is called a *knight move path*.

In a 3×3 square, where the fields have to be numbered from 1 to 9, such a knight move path covering all cells does not exist since from the middle cell no knight move is possible. But how about a 4×4 square?

Challenge:

The challenge consists of two questions:

- Is it possible to number the cells of a 4×4 square from 1 to 16 such that the path from 1 to 16 forms a knight move path?
- Is it possible to number the cells of a 4×4 square from 1 to 16 such that the path from 1 to 15 forms a knight move path, and the two leftmost numbers on the top row add up to 31?

Anyone who is skilled in programming may write a program for this that solves it. In this chapter about programming, you might expect that that is the way to go. But that is not the case: it should be done by reasoning only, without using any computer or calculator. Good luck!

More complicated sequences

In previous chapters, we have seen periodic and ultimately periodic sequences, and all kinds of corresponding turtle figures. Roughly speaking, these turtle figures appear in two flavors: either it is a finite picture with a regular pattern, or it continues forever in one direction, again with a regular pattern. But what kinds of sequences do exist that are *not* ultimately periodic, and how do their turtle figures look like? That is what we want to find out in this chapter. Before we get to that, we give an instruction only for readers that already work on the last chapter's challenge: if B is even, then A is increased by 10, otherwise everything stays the same, and in both cases the process continues to the text just before the challenge on knight moves.

Random sequences

A sequence a being not ultimately periodic can be made by flipping a coin for all values $i = 0, 1, 2, 3, \ldots$, and define $a_i = 0$ if the result is heads, and $a_i = 1$ if it is tails. This is impossible to do for all infinitely many values of i, but it is approximated by doing it for the first n elements for a big number n, and then make the turtle figure of that for some given angles. This would be a hassle when doing this physically with a coin. Luckily, this is easily simulated by a program. In many programming languages, including Lazarus, there is a command `random` for doing so. More precisely, by executing the command

```
a[i] := random(2)
```

for all values i running from 0 through $n-1$, all those a_i values are randomly set to 0 or 1, just like flipping a coin. We will not discuss how such a *random generator* internally works, we just assume that it behaves like flipping a coin.

DOI: 10.1201/9781003466000-8

So roughly half the time the result will be 0 and the other half it will be 1. And a particular pattern, for instance 000000, will occur occasionally, but not very often. All such phenomena of flipping coins will similarly show up for the random generator.

When making the turtle figure of the resulting first n elements, no patterns at all are expected. But it is nice to see how this will look like. Let's make such an initial part of a random sequence. We define $h(0) = 90$ and $h(1) = -90$, and choose $n = 20,000$. If we do this a number of times, we get different turtle figures for these first $20,000$ elements that we chose at random. One of them is the following:

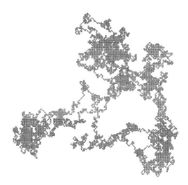

The starting point is at the bottom here, slightly left from the center. As the angles are straight, the turtle figure moves around in a rectangular grid. Some line segments are drawn several times, and of some pieces of the grid even all line segments are drawn. Due to this duplication of segements, the end point cannot be detected from the picture.

Next we do the same using other angles, let's pick at random: $h(0) = 120$ and $h(1) = -72$, and an even bigger initial part with $n = 100,000$. A resulting turtle figure for this initial part is then

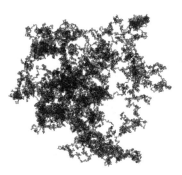

And here too, as expected, we see a very erratic pattern without recognizing any structures. On the internet I found something similar where the sequence was not made by a random generator, but the binary development of the number π was taken. The resulting turtle figure then looks just as random without any pattern.

The spiral sequence

Such a random sequence gives no patterns at all in the corresponding turtle figures. On the other hand, the ultimately periodic sequences we considered earlier only give very regular turtle figures. So more interesting sequences may contain some kind of structure, but are not ultimately periodic. A first corresponding sequence is the *spiral sequence*. Later we will see that this is an instance of the general class of morphic sequences.

The *spiral sequence* spir reads

$$\text{spir} = 1101001000100001 \cdots$$

More precisely, it is composed from infinitely many ones, with ever-growing groups of zeros in between: between the first two ones there are no zeros, between the second and third one there is one zero, between the third and fourth one there are two zeros, and so on. Here every subsequent group of zeros is exactly one greater than the previous one. The corresponding turtle figure with $h(0) = 0$ and $h(1) = 90$, so at 0 going straight ahead, and at 1 making a straight angle, looks like this:

Here the turtle figure starts in the middle in $(0,0)$, then goes up by the first 1 to $(0,1)$, then goes two steps to the left to $(-2,1)$ by processing 10. Then by processing 100 it moves down three steps to $(-2,-2)$, and so on. The part drawn here is only an initial part of around 2000 steps. The turtle figure of the full sequence spir continues indefinitely in all directions. Since this produces a spiral as we see, the sequence is called the *spiral sequence*.

It is also possible with angles other than straight angles: if we choose $h(0) = 0$ and $h(1) = 30$, an initial part of the turtle figure looks like this:

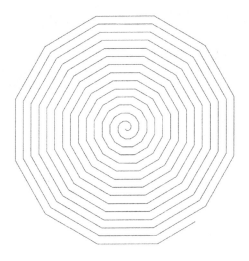

showing a spiral even more clearly. Choosing $h(0) = 0$ and $h(1) = 150$, the initial part of the first $20,000$ symbols gives the following turtle figure:

Until now, for the entire spiral sequence spir its turtle figures continue infinitely far in all directions. Surprisingly, this is no longer the case if we choose $h(0)$ to be non-zero. For example, if we choose $h(0) = 5$ and $h(1) = 108$, then after more than $300,000$ steps the turtle figure arrives at a point after which only line segments are drawn that were drawn before, just like with periodic sequences, and then the following turtle figure is obtained:

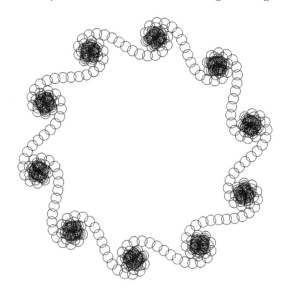

Hence this is the finite turtle figure of the entire infinite spiral sequence. Why this is the case, and why it takes more than $300,000$ steps, will be the challenge in Chapter 9 where we will see more finite turtle figures.

Pure morphic sequences

We are now going to describe a way to precisely define the spiral sequence and many other sequences. The notion *morphism* plays a central role. As a reminder, a morphism $f : A \rightarrow A^+$ is a function that maps every symbol s from A to a non-empty word $f(s)$. If u is a word, then $f(u)$ is the word obtained by replacing every symbol s in u by the word $f(s)$. If a is an infinite sequence then $f(a)$ is the infinite sequence obtained by replacing every symbol s in a by the word $f(s)$. Let's see what happens to spir if we apply the morphism f defined by

$$f(0) = 0, \quad f(1) = 01.$$

Applying this morphism f to a sequence composed from zeros and ones means that every 0 remains unchanged and in front of every 1 an extra 0 is added. Doing this for the spiral sequence spir yields

$$f(\text{spir}) \; = \; 0101001000100001 \cdots$$

Note that this is nearly the same sequence as spir, with the only difference that the first symbol 1 has been replaced by 0.

What we would like for *morphic sequences* is that it is a *fixed point* of a morphism f, i.e.

$$f(a) = a.$$

In words, a sequence is a fixed point of a morphism f if f applied to that sequence yields the same sequence again.

It sounds a bit strange to call a sequence a *point*. However, for all kinds of operations f such as reflections and rotations in the plane one may wonder for which points x the property $f(x) = x$ holds. For a reflection, these are all points on the line that are mirrored against, and for a rotation, that is the point around which the rotation is performed. Such points are called *fixed points*. Next, the word fixed point became a standard notion in mathematics, and is also used for operations on different objects than points in a plane or space. As the requirement $f(x) = x$ is exactly the same, it is natural also to use it for morphisms on sequences.

Back to our spiral sequence spir. That is now almost a fixed point of the morphism f, but not quite yet, because the first element has been changed. But this is easily fixed as follows. We define spir$'$ to be a copy of spir in which the first 1 is replaced by 2, so

$$\text{spir}' \; = \; 2101001000100001 \cdots$$

Now choosing $f(0) = 0$, $f(1) = 01$, $f(2) = 21$ yields

$$f(\text{spir}') \; = \; \underbrace{21}_{f(2)} \; \underbrace{01}_{f(1)} \; \underbrace{0}_{f(0)} \; \underbrace{01}_{f(1)} \; \underbrace{0}_{f(0)} \cdots \; = \; \text{spir}'.$$

So the sequence spir$'$ is indeed a fixed point of f.

Now we arrive at a very important observation: spir$'$ is not just any fixed point of f, it is the *only* fixed point of f that starts with the symbol 2. Let's see why. Let a be any fixed point of f that starts with 2. Since a starts with 2, $f(a)$ starts with $f(2) = 21$. But since $f(a) = a$, then a starts with 21. But then $a = f(a)$ starts with $f(21) = 2101$. This argument continues: a starts with

$f(2101) = 2101001$, also with $f(2101001) = 21010010001$, and so on. So for any finite initial part of spir', this is also an initial part of a. Hence a equals spir'.

This argument is given for the particular sequence spir', but the following statement shows that it is much more general. We use the notation f^n to apply n times f, so $f^1(s) = f(s)$ and $f^{n+1}(s) = f(f^n(s))$ for every s and every $n > 0$.

Theorem 8.1 *For a morphism $f : A \rightarrow A^+$ for which $f(s) = su$ for a symbol s in A and a non-empty word u there is exactly one sequence that starts with s and is a fixed point of f, which can be written as*

$$f^\infty(s) = suf(u)f^2(u)f^3(u)f^4(u)\cdots$$

This theorem introduces the new notation $f^\infty(s)$. Choosing this notation is motivated by the fact that from $f(s) = su$ it follows:

$$
\begin{aligned}
f^1(s) &= f(s) = su, \\
f^2(s) &= f(f(s)) = f(su) = f(s)f(u) = suf(u), \\
f^3(s) &= f(f^2(s)) = f(suf(u)) = suf(u)f(f(u)) = suf(u)f^2(u), \\
f^4(s) &= f(f^3(s)) = f(suf(u)f(f(u))) = suf(u)f^2(u)f^3(u), \\
&\cdots
\end{aligned}
$$

Now we give the proof of Theorem 8.1.

Proof: First note that $f^\infty(s) = suf(u)f^2(u)f^3(u)f^4(u)\cdots$ is indeed an infinite sequence: since u is non-empty and f maps any symbol to a non-empty word, $u, f(u), f^2(u), f^3(u), f^4(u), \ldots$ are all non-empty words, so the result is infinitely long.

We then establish that $f^\infty(s)$ is indeed a fixed point of f:

$$
\begin{aligned}
f(f^\infty(s)) &= f(suf(u)f^2(u)f^3(u)f^4(u)\ldots) \\
&= f(s)f(u)f(f(u))f(f^2(u))f(f^3(u))\ldots \\
&= suf(u)f^2(u)f^3(u)f^4(u)\ldots \\
&= f^\infty(s).
\end{aligned}
$$

It remains to show that $f^\infty(s)$ is the only fixed point of f that starts with s. Suppose a is any fixed point of f starting with s, then we have to show that $a = f^\infty(s)$. To do this we have to show that for every $n \geq 0$ we have $a_n = (f^\infty(s))_n$.

Since a starts with s, $f(a)$ starts with $f(s) = su$. Since a is a fixed point of f, $f(a) = a$, we conclude that a starts with $f(s) = su$. But because a starts with su, $a = f(a)$ starts with $f(su) = suf(u)$. This argument may be repeated any number of times. For every n there is a k such that the length of $suf(u)f^2(u)f^3(u)\ldots f^k(u)$ is at least n. By repeating the above argument k times, we see that the sequence a starts with $suf(u)f^2(u)f^3(u)\ldots f^k(u)$, and since this covers a_n on position n, we have $a_n = (f^\infty(s))_n$. This holds for every n, so $a = f^\infty(s)$, concluding the proof of the theorem. $\square$

Turtle figures of sequences of this particular form $f^\infty(s)$ are the main topic of the rest of this book. Some of them are finite, but show much more patterns than the turtle figures of ultimately periodic sequences that we saw before. Others are *fractal*, and are therefore infinite, and show special patterns. But before going into further detail, some terminology is useful.

Definition 8.1 A sequence a with symbols from a set A is called *purely morphic* over A if there is a morphism $f : A \to A^+$ such that $f(s) = su$ for a symbol s in A and a non-empty word u, and $a = f^\infty(s)$.

According to Theorem 8.1, this is a fixed point of f: $f(a) = a$. Hence spir$'$ is an example of a purely morphic sequence.

Morphic sequences

We presented the spiral sequence spir, obtained from spir$'$ by replacing 2 with 1. Such a replacement of symbols is called a *coding*. In this case, we can say that the coding c is a function $c : A \to B$. Such a code is applied to individual elements of A, but is also defined on words and sequences over A. In the case of spir the set A consists of $0,1$ and 2, and B consists of 0 and 1, and $c : A \to B$ is defined by $c(0) = 0$, $c(1) = 1$, $c(2) = 1$.

Definition 8.2 A sequence b over B is called *morphic* if $b = c(a)$ for a purely morphic sequence a over some alphabet A and a coding $c : A \to B$.

Hence spir is an example of a morphic sequence. Every purely morphic sequence is also morphic: we may choose $A = B$, and let c maps every symbol to itself.

How about the other direction, is there really a difference between morphic and pure morphic sequences? By construction, spir is morphic. Now we will show that spir *not* is purely morphic, so there is a real difference between these two concepts.

Suppose that spir is purely morphic, then A consists of the two symbols 0 and 1, and spir $= f^\infty(s)$ for s with $f(s) = su$. Because spir starts with 1, we have $s = 1$. So the morphism f is fixed by $f(0) = v$, $f(1) = 1u$ for certain non-empty words u and v. From the assumption

$$\text{spir} = 11010010001 \cdots = f^\infty(1) = 1uf(u)f^2(u)f^3(u) \cdots$$

we see that u starts with 1. But then there are two consecutive ones in $f(1) = 1u$, and because u contains at least one 1, there are also two consecutive ones in $f(u)$, being part of spir $= f^\infty(1)$ which is not at the beginning. But other than the first two ones, there are no two consecutive ones in spir anywhere, so that's a contradiction. This proves that spir is not purely morphic.

Of course turtle figures of morphic sequences may be considered instead of purely morphic sequences. But that will give nothing new. Namely, if we have a turtle figure for a morphic sequence $c(f^\infty(s))$ over B where $f : A \to A^+$ and $c : A \to B$, and the angles $h(s)$ are defined for symbols s from B, then we get exactly the same turtle figure for the purely morphic sequence $f^\infty(s)$ by defining $h(s) = h(c(s))$ for all symbols s from A. Hence for considering turtle figure of morphic sequences, we may and will restrict to pure morphic sequences. For instance, for our spir example, we get the same turtle figure as for spir$'$ if we define $h(2)$ to be equal to the angle $h(1)$.

While we're discussing angles, it's a good time now to consider the challenge on subwords. We had a word u of length 360, and for every symbol s the angle $h(s)$ is an integer number of degrees. For each i from 0 to 360 we define $h_i = h(w)$ where w consists of the first i symbols of u. So $h_0 = 0$ and $h_{360} = h(u)$. Since for every s, the angle $h(s)$ is an integer, the angle $h_i = h(w)$ is an integer too. So for all 361 values i from 0 through 360, the angle h_i is an integer number of degrees, for which we have exactly 360 possibilities, being less then 361. Hence, there are two distinct values $i < j$ for which $h_i = h_j$. Write w for the word consisting of the first i symbols of u. The word consisting of the first j symbols of u starts with w, so we can write that word as wv. Since $i < j$ the word v is non-empty. And because $h(w) = h_i = h_j = h(wv) = h(w) + h(v)$ we conclude $h(v) = 0$. So the claim stated in the challenge is true.

We want to give two general comments here. First, such a counting argument is quite common: from the fact that more than n elements are chosen from exactly n possibilities, it is concluded that at least one of these n possibilities is chosen at least twice. This principle is called the *pigeonhole principle*. Here the word *pigeon-hole* does not only refer to the pigeon bird, but also to a *mailbox*. The corresponding story is that if you have more letters than mailboxes, there is at least one mailbox with more than one letter, if all the letters are put in the boxes. This obvious principle is sometimes attributed to Dirichlet (1805–1859), but was also applied before that.

The second comment is about the fact that in our argument we did not use part of the given information, namely that the alphabet consists of seven symbols. The result we derived above holds for any number of symbols, even if all 360 symbols are different and all have a different angle. In a way, an exercise may become more difficult if it contains redundant information like here, due to the resulting confusion. When formulating a theorem, it is a general principle to present the theorem as strong as possible, and only include information that is really needed in giving the proof.

Programming morphic sequences

We now show how to compute an initial part of a purely morphic sequence, and how to generate it by a Python program. Such a purely morphic sequence is of the shape

$$f^\infty(s) = suf(u)f^2(u)f^3(u)f^4(u)\ldots,$$

and thus starts with su. In the remaining part

$$f(u)f^2(u)f^3(u)f^4(u)\ldots$$

only symbols appear as part of a word $f(s')$, where s' is some symbol earlier in the sequence. And these symbols s' are processed one by one.

As an example we take the sequence $a = f^\infty(0)$ for f defined by $f(0) = 01$ and $f(1) = 0$. We will later call this the *binary Fibonacci sequence* and discuss it in more detail. Now we show how to construct an initial part of any length of this sequence. We start by $f(0) = 01$. At this point we only have 01, which is the f applied on the first symbol, being 0. Now the sequence continues with f applied on the second symbol. Since we start counting at 0 (the first element of the sequence is a_0), it means that for $i = 1, 2, 3, \ldots$ we have to apply f on a_i and then add $f(a_i)$ at the end of the initial part constructed so far. Since for every symbol s the word $f(s)$ is non-empty and thus has length ≥ 1, these a_i are always defined before you need them. This may be done by hand: after the 01 we add $f(1) = 0$, then $f(0) = 01$, and so on: $a = f^\infty(0) = 0100101001001 \cdots$

This is expressed in a Python program as follows. Starting with 01 can be written in Python as `a = [0,1]`.

Then for $i = 1, 2, 3, \ldots$ we need to add the word $f(a_i)$. We do that by testing using an `if` command whether a_i is 0: if so, we should add $f(0) = 01$, otherwise $f(1) = 0$. In Python, we may use the command `append` to append elements. The resulting program will then look like this:

```
a = [0,1]
for i in range(n):
    if a[i+1] == 0:
        a.append(0)
        a.append(1)
    else:
        a.append(0)
```

Here the value of n has to be chosen, depending on the desired size of the initial part of the sequence a.

If we then want to draw the turtle figure of this initial part of a, this may be done by extending the program by going through the values a[i] using `for i in range`, repeating rotating the corresponding angle and taking a step.

The full program might look like this:

```
from turtle import *
speed(0)
a = [0,1]
d = 80
h0 = 144
h1 = 36
for i in range(2500):
    if a[i+1] == 0:
        a.append(0)
        a.append(1)
    else:
        a.append(0)
for i in range(2500):
    if a[i] == 0:
        left(h0)
    else:
        left(h1)
    forward(d)
done()
```

Executing this progam yields the corresponding turtle figure:

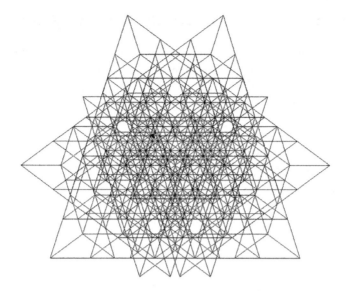

This is just an example. In exactly the same way, for every purely morphic sequence (and therefore every morphic sequence) $f^\infty(s)$ the turtle figure of any finite initial part may be created: start by giving a the value $f(s)$, give the angles h the desired values, and in the first part of the program replace the **appends** by what the f definition prescribes. This is the way all turtle figures of morphic sequences in this book are created, in most cases elaborated in Lazarus instead of Python.

For the reader that may use a hint for the knigth move challenge: the 16 squares on the 4×4 square may be divided into three types:

- the four corners, denoted by a,
- the eight other border cells, denoted by b, and
- the remaining four inside cells, denoted by c.

Below this is shown in a picture. Next one may figure out which knight moves are possible from which letters to which letters. That will give the key to the solution as we will give later.

a	b	b	a
b	c	c	b
b	c	c	b
a	b	b	a

Challenge: a variation on the spiral sequence

The spiral sequence is also denoted as $\langle n \rangle$, where the n represents the fact that the sizes of the groups of consecutive between the ones are $0, 1, 2, 3, \ldots$ Similarly one writes

$$\langle 2^n \rangle = 10100100001000000001 \cdots$$

in which the sizes of the groups of consecutive zeros are $1, 2, 4, 8, 16, \ldots$, the powers of 2. This is given as a morphic sequence by $\langle 2^n \rangle = c(f^\infty(2))$ for the morphism f and the coding c defined by

$$f(0) = 00, \; f(1) = 1, \; f(2) = 201, \; c(0) = 0, \; c(1) = 1, \; c(2) = 1.$$

Challenge:

Is it possible to present

$$\langle n^2 \rangle = 1101000010000000001 \cdots$$

as a morphic sequence? Here the sizes of the groups of consecutive zeros are $0, 1, 4, 9, 16, \ldots$, being the successive squares.

If it is possible: show how, and if not, indicate why not.

The Thue-Morse sequence

A fair distribution

Suppose we want to make an infinite sequence over zeros and ones that contain the zeros and ones distributed as evenly as possible: as many zeros as ones. Let's start by 0. Then the next element will be 1 for the desired fairness. Repeating this endlessly results in our first try

$$0101010101010\cdots = (01)^\infty.$$

Indeed, if you take the first hundred symbols of this, then exactly 50 are 0 and exactly 50 are 1, that looks very fair. Similarly, for any other even number $2n$ among the first $2n$ symbols exactly n of them are 0 and also exactly n are 1. This looks as fair as possible.

But what if you take an odd number? Of the first five, there are three zeros and two ones, that is is not entirely fair because three is more than two. This is inevitable because an odd number cannot be split into two equal parts, that's just a property of odd numbers. But what we do see is that for every odd number $2n + 1$, among the first $2n + 1$ elements there are exactly $n + 1$ equal to 0 and exactly n equal to 1. Hence, there are always more zeros than ones in any initial part of the sequence of odd length, and never the other way around. So for these odd length initial parts, the division between zeros and ones is not fair.

Is it possible to fix this? Let's consider the following second attempt:

$$011001100110\cdots = (0110)^\infty.$$

Now again among the first $2n$ symbols of this sequence there are exactly n equal to 0 and also exactly n equal to 1, so completely fair, just like in the

DOI: 10.1201/9781003466000-9

first example. But now for odd length initial parts, we alternately obtain a surplus of ones and zeros: for the first 3, 7, 11 symbols of this sequence, its number of ones is exactly one more than the number of zeros, while of the first 1, 5, 9, 13 symbols the number of zeros is exactly one more than the number of ones. So this is much more fair than the first example.

Are we completely satisfied now? If in a sequence the distribution of zeros and ones is completely fair, one may hope that this remains true after removing elements from the sequence in some way. Of course if you remove all zeros or all ones, the result will not have a fair distribution. But if you remove elements in a non-biased way, you may hope that the fair distribution remains. One way to do this is by alternately keeping and removing a symbol. Doing so for a sequence $a = a_0a_1a_2a_3\cdots$ then yields the sequence

$$\text{even}(a) = a_0a_2a_4a_6\cdots \text{ and } \text{odd}(a) = a_1a_3a_5a_7\cdots.$$

Now we try to choose our sequence in such a way that it not only has a fair distribution of zeros and ones itself, but also if you apply even or odd on it. Let's see what happens to the sequence that we considered so far. For the first example, we obtain $\text{even}((01)^\infty) = 0000\cdots = 0^\infty$ and $\text{odd}((01)^\infty) = 1111\cdots = 1^\infty$: both having an extremely unfair distribution of zeros and ones. For our second attempt $(0110)^\infty$ we obtain $\text{even}((0110)^\infty) = 010101\cdots = (01)^\infty$ and $\text{odd}((0110)^\infty) = 101010\cdots = (10)^\infty$. For $(01)^\infty$, we already argued that it doesn't have a completely fair distribution of zeros and ones, and a similarly argument holds for $(10)^\infty$.

Is it possible to repair this? A next attempt would be $a = (01101001)^\infty$. But then applying even once more will yield an unfair distribution, for instance, we obtain $\text{even}(\text{even}(\text{even}(a))) = 0^\infty$, not having a fair distribution at all. Would it be possible to construct a sequence, such that, no matter how often we apply odd or even, the result always has a fair distribution over zeros and ones?

The Thue-Morse sequence as a morphic sequence

Fortunately, our notion of morphic sequences will provide a solution. If we continue the process just sketched indefinitely, we end up by the purely morphic sequence

$$\mathbf{t} = f^\infty(0) = 0110100110010110\cdots$$

for the morphism f defined by $f(0) = 01$ and $f(1) = 10$.

Since $\mathbf{t} = f(\mathbf{t}) = f(0)f(1)f(1)f(0) \cdots$, applying even on this sequence removes every second element of every $f(0) = 01$ and $f(1) = 10$, leaving only 0 from $f(0)$ and 1 from $f(1)$, resulting in even($\mathbf{t}$) = $\mathbf{t}$. So no matter how many times we apply even to $\mathbf{t}$, the result remains $\mathbf{t}$ in which the zeros and ones are evenly distributed.

Let's write $\mathbf{t}' = 10010110 \cdots$ for the sequence obtained by replacing every 0 by 1 and every 1 by 0 in $\mathbf{t}$. Then we have $\mathbf{t}' = f^\infty(1)$. Due to the swapping of 0 and 1, the sequence $\mathbf{t}'$ has the same fair distribution as $\mathbf{t}$. Since swapping 0 and 1 in $\mathbf{t} = f(\mathbf{t})$ is the same as taking out the first element of every $f(0) = 01$ and $f(1) = 10$, we obtain odd($\mathbf{t}$) = $\mathbf{t}'$, and similarly odd($\mathbf{t}'$) = $\mathbf{t}$. So we get an even stronger result: no matter how many times we apply even and odd on $\mathbf{t}$ or $\mathbf{t}'$, in any order, the result is always $\mathbf{t}$ or $\mathbf{t}'$, in which the zeros and ones are evenly distributed.

Alternative characterizations

This sequence $\mathbf{t}$ has several remarkable properties. A way to determine element $\mathbf{t}_n$ of $\mathbf{t} = \mathbf{t}_0\mathbf{t}_1\mathbf{t}_2 \cdots$ without any knowledge of morphic sequences is as follows. Take the *binary notation* of n. If the number of ones in it is even, then $\mathbf{t}_n = 0$, and if it is odd, then $\mathbf{t}_n = 1$. Let's check this for some small values of n. The binary notation of 0 is 0, or empty, with zero zeros, zero is even, so $\mathbf{t}_0 = 0$. The next two elements are $\mathbf{t}_1$ and $\mathbf{t}_2$, the binary notations of 1 and 2 are 1 and 10, each having one symbol 1, so an odd number of ones, so $\mathbf{t}_1 = \mathbf{t}_2 = 1$. The number 3 is 11 in binary notation, having an even number of ones, so $\mathbf{t}_3 = 0$. And so on. Let's use this to establish $\mathbf{t}_n$ for some larger n, say $n = 83$. Writing $83 = 64 + 16 + 2 + 1$, we see that the binary notation of 83 is 1010011, having four ones, an even number, so $\mathbf{t}_{83} = 0$.

Correctness of this remarkable property follows from the next characterization of $\mathbf{t}$. From the property even($\mathbf{t}$) = $\mathbf{t}$ we conclude that for every i we have $\mathbf{t}_{2i} = (\text{even}(\mathbf{t}))_i = \mathbf{t}_i$. From the property odd($\mathbf{t}$) = $\mathbf{t}'$ we conclude that for every i we have $\mathbf{t}_{2i+1} = (\text{odd}(\mathbf{t}))_i = \mathbf{t}'_i = \neg\mathbf{t}_i$. Here $\neg$ is the operation that swaps the 0 and 1, being defined by $\neg i = 1 - i$ for i equal to 0 or 1. In summary, we have

$$\begin{aligned}
\mathbf{t}_0 &= 0, \\
\mathbf{t}_{2i} &= \mathbf{t}_i \quad \text{for each } i, \\
\mathbf{t}_{2i+1} &= \neg\mathbf{t}_i \quad \text{for each } i.
\end{aligned}$$

Now we observe that t_i is completely defined for every $i \geq 0$ by these three equations. The equations may be seen as an inductive definition of $\mathbf{t}$: first t_0 is defined, and next for every $i > 0$, the intended value of t_i expressed in t_j for some $j < i$. As an example, we again determine the value of t_{83}, now using these equations:

$$\mathbf{t}_{83} = \neg \mathbf{t}_{41} = \neg\neg \mathbf{t}_{20} = \mathbf{t}_{20} = \mathbf{t}_{10} = \mathbf{t}_5 = \neg \mathbf{t}_2 = \neg \mathbf{t}_1 = \neg\neg \mathbf{t}_0 = \mathbf{t}_0 = 0.$$

What is going on here is essentially the same as counting the ones in binary notation, processing the binary digits from right to left. An even number in binary notation ends in 0. By the second equation this is removed and has no influence on counting ones. An odd number in binary notation ends in 1. By the last equation this is removed and yields $\neg$ in order to swap the parity of counting ones.

This sequence $\mathbf{t}$ has been discovered and studied independently by several people. Two of them are the Norwegian Axel Thue (1863–1922) and the American Marston Morse (1892–1977). The latter is not the one from Morse code: that was Samuel Morse who lived a hundred years earlier. A common name for the sequence $\mathbf{t}$ is therefore *Thue-Morse sequence*, which is why it is denoted by the letter $\mathbf{t}$. In fact, it was discovered and studied much earlier by the Frenchman Eugène Prouhet (1817–1867). A remarkable property is that $\mathbf{t}$ is *cube free*, that is, does not contain the pattern *uuu* for a non-empty word *u*. In a later section in this chapter we will prove this property. In fact, $\mathbf{t}$ is one of the simplest sequences that is cube free. One more person that independently discovered this sequence was Max Euwe (1901–1981) from the Netherlands, who became famous by being the world champion in chess from 1935 to 1937. In chess a rule states that it is a draw if exactly the same position occurs three times in the game. Euwe used the cube freeness of the Thue-Morse sequence to prove that a draw may always be prevented in a particular case. In an interview Euwe once was asked what he was most proud of in his life. Of course everyone expected that this was his world championship. But he answered differently: he was most proud of the fact that he was appointed professor of computer science shortly before his retirement.

Finite turtle figures of more general sequences

When considering periodic sequences we saw in Theorem 6.1 that their turtle figures are often finite. We saw many examples of them. The underlying idea was that if the angle $h(u)$ is rational, then the angle $h(u^n)$ equals 0 for some n.

As a consequence we were able to prove that often the position $P(u^n)$ is equal to $(0,0)$. Combined with $h(u^n) = 0$, this shows that after processing u^n the turtle figure is back where it started, and in the rest of the turtle figure of $u^\infty = (u^n)^\infty$ only lines segments are drawn that were drawn before. We will now show how a similar argument for morphic sequences shows finiteness of turtle figures under certain conditions. We focus on the case where the alphabet consists of two symbols 0 and 1, but the properties we show are easily generalized to larger alphabets.

The crux of the argument that the turtle figure of a periodic sequence is finite lies in the fact that the sequence can be written as u^∞ for a word u for which both the angle $h(u) = 0$ and the position $P(u) = (0,0)$ are equal to their original value. In the above story that is the case for u^n. More general, the same argument holds for a sequence a that is composed from non-empty words u and v, in any order, satisfying $h(u) = h(v) = 0$ and $P(u) = P(v) = (0,0)$. Each line segment of the turtle figure of a is then a line segment of the turtle figure of either u or v. The total number of these line segments is therefore at most the length of u plus the length of v.

The set of sequences composed from u and v consecutively, in any order, is denoted by $\{u,v\}^\infty$. For example, u^∞, v^∞ and $(uv)^\infty$ are all sequences contained in $\{u,v\}^\infty$. More general: if a is any sequence over the symbols 0 and 1, and we define the morphism f by $f(0) = u$ and $f(1) = v$, then $f(a)$ is a sequence contained in $\{u,v\}^\infty$. This shows the following property:

Theorem 9.1 *Let u and v be non-empty words of lengths n and m, satisfying $h(u) = h(v) = 0$ and $P(u) = P(v) = (0,0)$. Let a be a sequence in $\{u,v\}^\infty$. Then the turtle figure of a is finite and consists of at most $n+m$ line segments.*

Finite turtle figures of the Thue-Morse sequence

The following turtle figure of the Thue-Morse sequence **t** is finite and consists of 128 line segments; the chosen angles are $h(0) = -132\frac{1}{2}$ and $h(1) = 20$.

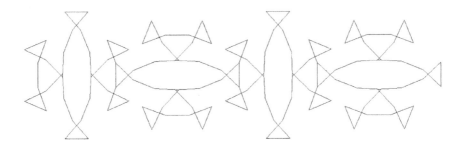

$$f(0) = 01, f(1) = 10, h(0) = -132\frac{1}{2}, h(1) = 20$$

We will now find out why this is the case. In order to do so, we start by some general properties.

Note that Theorem 9.1 applies to general sequences, not limited to the Thue-Morse sequence, and not even to morphic sequences. Now we will show how it applies to morphic sequences, in particular to the Thue-Morse sequence. The key observation is the following:

Theorem 9.2 *Every purely morphic sequence* $a = f^\infty(s)$ *over 0 and 1 is contained in* $\{f^n(0), f^n(1)\}$*, for every natural number n.*

Proof: Take a natural number n. Since a is a fixed point of f we have

$$a = f(a) = f(f(a)) = f(f(f(a))) = \cdots = f^n(a).$$

Hence

$$a = f^n(a_0)f^n(a_1)f^n(a_2)f^n(a_3)\cdots$$

is in $\{f^n(0), f^n(1)\}$ since every a_i is either 0 or 1. $\square$

Indeed, for $f(0) = 01$ and $f(1) = 10$ we have $f^3(0) = 01101001$ and $f^3(1) = 10010110$, and splitting up **t** into blocks of length 8 yields blocks all being equal to either 01101001 or 10010110.

We now want to combine Theorems 9.1 and 9.2 to conclude that at certain angles $h(0)$ and $h(1)$ the sequence **t** yields a finite turtle figure. We have to choose $h(0)$ and $h(1)$ and a number n such that $h(f^n(0)) = h(f^n(1)) = 0$ and $P(f^n(0)) = P(f^n(1)) = (0,0)$.

Lemma 9.1 *Let* $f(0) = 01$ *and* $f(1) = 10$, *and* $h(0) + h(1) = \frac{360k}{2^n}$ *for an odd number k. Then*

$$h(f^{n+1}(0)) = h(f^{n+1}(1)) = 0$$

and

$$P(f^{n+2}(0)) = P(f^{n+2}(1)) = (0,0).$$

Proof: Both $f^{n+1}(0)$ and $f^{n+1}(1)$ consist of exactly 2^n zeros and 2^n ones. So $h(f^{n+1}(0)) = h(f^{n+1}(1)) = 2^n(h(0) + h(1)) = 360k$, being equal to 0 considered as an angle.

From Chapter 7 on turtle figures we use the notation $R_h(x, y)$ for the rotation of point (x, y) by an angle h. There we observed that

$$P(uv) = P(u) + R_{h(u)}(P(v)).$$

Some basic properties of this rotation were already collected in Lemma 6.2. Let's add one more: we always have $R_{180}(x, y) = -(x, y) = (-x, -y)$. This states that rotating over 180° is exactly the same as mirroring about the point $(0,0)$, which converts (x, y) to $-(x, y) = (-x, -y)$. Using this property we obtain: if $h(u) = 180$, then $P(uu) = P(u) - P(u) = (0,0)$. Now we choose $u = f^n(0)$ and $v = f^n(1)$. Since u and v each consist of exactly 2^{n-1} zeros and 2^{n-1} ones, we obtain $h(u) = h(v) = 2^{n-1}(h(0) + h(1)) = 2^{n-1}\frac{360k}{2^n} = 180k$. The odd number k times 180 degrees is again 180 degrees, so we have $h(u) = h(v) = 180$. Applying the observations now yields $P(uu) = P(vv) = 0$. Moreover, we have $f^{n+2}(0) = f^n(f^2(0)) = f^n(0110) = f^n(0)f^n(1)f^n(1)f^n(0) = uvvu$, yielding

$$
\begin{aligned}
P(f^{n+2}(0)) &= P(uvvu) = P(u) - P(vvu) \\
&= P(u) - (P(vv) + R_{h(vv)}P(u)) \\
&= P(u) - P(u) = (0,0).
\end{aligned}
$$

By swapping u and v in this argument we similarly obtain $P(f^{n+2}(1)) = (0,0)$, concluding the proof. $\square$

Now we exploit this lemma to conclude finiteness of turtle figures.

Theorem 9.3 *Let $h(0) + h(1) = \frac{360k}{2^n}$ for an odd number k. Then the corresponding turtle figure of* **t** *is finite, and consists of at most 2^{n+3} line segments.*

Proof: By definition we have **t** $= f^{\infty}(0)$ for the morphism f defined by $f(0) = 01$ and $f(1) = 10$. Note that $f^{n+2}(0) = f^{n+1}(01) = f^{n+1}(0)f^{n+1}(1)$ and $f^{n+2}(1) = f^{n+1}(10) = f^{n+1}(1)f^{n+1}(0)$, so $h(f^{n+2}(0)) = h(f^{n+2}(1)) = h(f^{n+1}(0)) + h(f^{n+1}(1))$. According to Lemma 9.1 we have $h(f^{n+1}(0)) = h(f^{n+1}(1)) = 0$, so also $h(f^{n+2}(0)) = h(f^{n+2}(1)) = 0$, and $P(f^{n+2}(0)) =$

$P(f^{n+2}(1)) = (0,0)$. For $u = f^{n+2}(0)$ and $v = f^{n+2}(1)$ we conclude from Theorem 9.2 that $\mathbf{t}$ is in $\{u,v\}^\infty$. Note that u and v both have length 2^{n+2}. Now we conclude from Theorem 9.1 that the turtle figure of $\mathbf{t}$ is finite and consists of at most $2^{n+2} + 2^{n+2} = 2^{n+3}$ line segments. $\square$

In the very first chapter we already saw an example: there we chose $h(0) = -146.25 = \frac{-13 \times 360}{2^5}$, and $h(1) = 4.74609375 = \frac{27 \times 360}{2^{11}}$, yielding $h(0) + h(1) = \frac{27 - 13 \times 64}{2^{11}}$. So the conditions of Theorem 9.3 are satisfied for $n = 11$, concluding that the turtle figure consists of at most $2^{14} = 16,384$ line segments.

In the example at the beginning of this section we had $h(0) = -132\frac{1}{2}$ and $h(1) = 20$, so $h(0) + h(1) = -112\frac{1}{2} = \frac{-5 \times 360}{2^4}$. Indeed by Theorem 9.3 we conclude that the resulting finite turtle figure has at most $2^7 = 128$ line segments. It turns out to be exactly 128.

Choosing $h(0) = \frac{360}{16}$ and $h(1) = \frac{63 \times 360}{128}$ gives the following turtle figure for $\mathbf{t}$ of 1024 line segments:

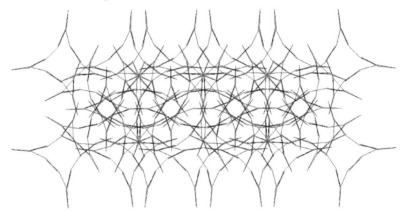

Again choosing $h(1) = \frac{63 \times 360}{128}$, but now $h(0) = \frac{3 \times 360}{32}$ gives the following turtle figure for $\mathbf{t}$ of 1024 line segments again:

Slightly more line segments are obtained for turtle figure for $\mathbf{t}$ by choosing $h(0) = \frac{3 \times 360}{32}$ and $h(1) = \frac{117 \times 360}{256}$, namely 2048:

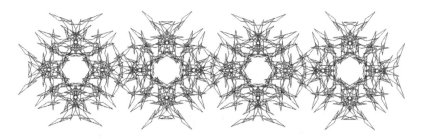

In all of these angles, we see powers of 2 in the denominator and an odd number times 360 in the numerator. These odd numbers were chosen quite randomly. In fact, I just tried several odd numbers, and the one odd number gave a more beautiful picture than the other. This notion 'beautiful' is of course very subjective, but a reasonable criterion seems to be some balance between order and arbitrariness. Only order is boring, and only randomness becomes a mess. A mixture of both seems to give the most exciting effect. Anyway, when playing around with several values, I was surprised by the turtle figures of the Thue-Morse sequence obtained by choosing angles that meet the conditions of Theorem 9.3. We continue by a few more examples having even more line segments.

The next one has $2^{14} = 16,384$ segments, just like the example in the first chapter, but now choosing $h(0) = \frac{61 \times 360}{128}$ and $h(1) = \frac{33 \times 360}{2048}$:

This is not yet the end. As the last of this series we give the turtle figure of **t** for $h(0) = 180$ and $h(1) = \frac{1097 \times 360}{8192}$, consisting of $2^{16} = 65,536$ segments:

Thue-Morse is cube free

After having seen all these finite turtle figures for the Thue-Morse sequence **t**, we now present the theory of some remarkable property of **t**: it does not contain a non-empty *cube*. Here a *cube* is defined to be a word consisting of three times the same subword. So a word w is a cube if it can be written as uuu for some word u. A word or sequence is called *cube free* if it does not contain a non-empty cube. So a sequence a is cube free if it can not be written as $a = vuuub$ for a word v, a non-empty word u and some sequence b. In this section we will show that **t** is cube free. In fact, **t** can be seen as one of the simplest cube free sequences over 0 and 1, and this was one of the reasons why this sequence was discovered independently several times. We will show an even stronger property: we will show that **t** does not contain an *overlap*. Here an overlap is defined to be a word of the shape $sxsxs$ where s is a single symbol and x is a word. Indeed, having no overlap is a stronger property than being cube free, since for every non-empty cube uuu the word u is non-empty and can be written as $u = sx$, where s is the first symbol of u and x is the rest, and the cube $uuu = sxsxsx$ contains the overlap $sxsxs$. As before we have $f(0) = 01, f(1) = 10$, $\mathbf{t} = f^\infty(0)$, $\mathbf{t}' = f^\infty(1)$. We recall that $\text{even}(\mathbf{t}) = \mathbf{t}$, $\text{odd}(\mathbf{t}) = \mathbf{t}'$, $\text{even}(\mathbf{t}') = \mathbf{t}'$, $\text{odd}(\mathbf{t}') = \mathbf{t}$, and by swapping 0 and 1 the sequences **t** and **t**′ transform to each other. The sequence **t** satisfies $\mathbf{t}_{2i} = \mathbf{t}_i$ and $\mathbf{t}_{2i+1} = \neg \mathbf{t}_i$ for every $i \geq 0$. These properties are sufficient for giving the proof, that needs to consider quite some case analysis on lengths of words being odd or even.

Theorem 9.4 *The sequence* **t** *contains no overlap.*

Proof: Assume that **t** contains an overlap. Then write $\mathbf{t} = vsxsxsa$ for some word v, some symbol s and some sequence a, and $sxsxs$ being a shortest possible overlap occurring in **t**. Write $x = x_0 x_1 \cdots x_{k-1}$ for $k = |x| \geq 0$ being the length of x. To obtain a contradiction, we distinguish the cases where k is either even or odd.

First let k be even. Write $k = 2m$. Write $|v| = p$, then from $\mathbf{t} = vsxsxsa$ we conclude that $x_i = \mathbf{t}_{p+i+1} = \mathbf{t}_{p+i+2m+2}$ for $i = 0, \ldots, k$. Since for every i either $p+i+1$ or $p+i+2m+2$ is even, from $\mathbf{t}_{2j+1} = \neg \mathbf{t}_j = \neg \mathbf{t}_{2j}$ for every j we obtain $x_{j+1} = \neg x_j$ for $j = 0, \ldots, k-1$. So either $x = (01)^m$ or $x = (10)^m$. We derive the contradiction for $x = (01)^m$, the argument for $x = (10)^m$ is similar. So $\mathbf{t} = vs(01)^m s(01)^m sa$. If p is even we have $\mathbf{t}_p = s$, $\mathbf{t}_{p+1} = 0$, $\mathbf{t}_{p+2m} = 1$ and $\mathbf{t}_{p+2m+1} = s$, then applying $\mathbf{t}_{p+1} = \neg \mathbf{t}_p$ and $\mathbf{t}_{p+2m+1} = \neg \mathbf{t}_{p+2m}$ yields

$0 = \neg s = 1$, contradiction. In the remaining case, p is odd and we have $t_{p+2m+1} = s$, $t_{p+2m+2} = 0$, $t_{p+4m+1} = 1$ and $t_{p+4m+2} = s$, then applying $t_{p+2m+2} = \neg t_{p+2m+1}$ and $t_{p+4m+2} = \neg t_{p+4m+1}$ again yields $0 = \neg s = 1$, contradiction. This argument also holds for $k = 0$, so we are done for k even.

In the remaining case k is odd. Hence sx has even length. In case v has also even length, we obtain $t = \text{even}(t) = \text{even}(v)\text{even}(sx)\text{even}(sx)\text{even}(sa)$, where we apply even on even length words with the obvious meaning. But now we write $\text{even}(sx) = su$ for a word u being shorter than x. But then t contains the overlap $susus$ being shorter than $sxsxs$, contradicting the assumption that $sxsxs$ is a shortest possible overlap in t. It remains to consider the case where the length of v is odd. Then $t' = \text{odd}(t) = \text{odd}(vsxsxsa) = \text{odd}(v)\text{even}(sx)\text{even}(sx)\text{even}(sa)$, where odd applied on the odd length word v means applying even after removing the first element. Similar as above we now obtain that t' contains a shorter overlap $susus$ than $sxsxs$. Now swapping every 0 and 1 t' yields t, containing the overlap obtained by applying this swap on $susus$. This is still an overlap shorter than $sxsxs$, contradicting the assumption that $sxsxs$ was the shortest overlap occurring in t. This concludes the proof. $\square$

A natural property being stronger than cube free and having no overlap is being *square free*, that is, does not contain the pattern uu for a non-empty word u. Square free sequences over two symbols 0 and 1 do not exist, since every word of length 4 either contains two consecutive equal symbols (hence being a square), or is a square itself: 0101 or 1010. But square free sequences over three or more symbols do exist. For instance, it can be shown that the pure morphic sequence $f^\infty(0)$ for $f(0) = 01201$, $f(1) = 020121$ and $f(2) = 0212021$, is square free. This example is taken from an exercise in the excellent book on automatic sequences by Allouche and Shallit, and we will say some more about this in the concluding chapter.

Stuttering variants of Thue-Morse

After this theoretical intermezzo let's come back to turtle figures of which finiteness is obtained by Theorem 9.1. This theorem can be applied to other sequences rather than only the Thue-Morse sequence. Here we apply it to two examples of variants of the Thue-Morse sequence that we call *stuttering*: these are of the shape $f^\infty(0)$ where 0 is not mapped to 01 but to a few copies of 01, and similarly for 1. Details of how to apply Theorem 9.1 for these examples are omitted, since the argument is very similar to the argument for the Thue-Morse sequence.

In the first example, we take two copies: we consider the sequence $f^\infty(0)$ for the morphism f defined by

$$f(0) = 0101, \quad f(1) = 1010.$$

As angles we choose $h(0) = \frac{43 \times 360}{128}$ and $h(1) = \frac{-3 \times 360}{256}$. The turtle figure of $f^\infty(0)$ is as follows:

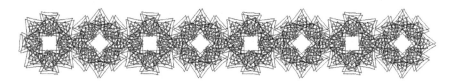

In the Thue-Morse sequence, we often saw a pattern repeated four times because of the stuttering it seems to have become eight here.

It will be more exciting if we take three copies, so

$$f(0) = 010101, \quad f(1) = 101010.$$

As angles we choose $h(0) = -64\frac{4}{9}$ and $h(1) = 135$, yielding the following turtle figure with threefold symmetry:

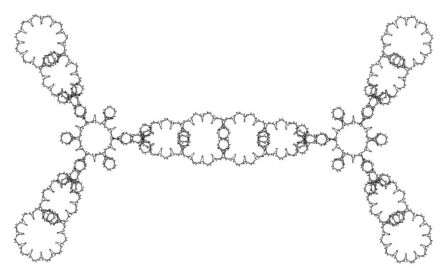

This turtle figure is obtained for the initial part of the sequence consisting of $2592 = 2^5 \times 3^4$ symbols. If we take half of this $= 1296$, we get only the right half of this figure, being completely threefold symmetric, that is, maps exactly to itself by rotating over 120°.

The end of this chapter on the Thue-Morse sequence, and these variants is approaching. Before giving the final challenge, we give the solution of an earlier challenge on knight moves. Each of the 16 cells in the square is marked by a, b, or c as follows:

a	b	b	a
b	c	c	b
b	c	c	b
a	b	b	a

About knight move paths, the following is observed:

- every a is followed and preceded by c,
- no c is followed or preceded by c,
- the pattern $cacac$ does not occur, because after $caca$ the only c you can get to is the c you started in, and
- more than 4 b's in a row is not possible, because after $bbbb$ the only b you can go to is the b you started with.

Assume a knight move path of length 16 squares visiting all squares exactly once. Then there are exactly 4 a's, exactly 8 b's and exactly 4 c's. According to the latest observation, the 8 bs are not consecutive, so we have at least two groups of b's. Only a single c or the pattern cac may occur between two groups of bs, since anything else is not possible according to the first three observations.

In front of the first group of b's only ac or $acac$ may occur, or something with more cs than as, and after the last group of b's can only ca or $caca$ may occur, or something with more c's than a's. Conclusion: in any path of length 16 that satisfies the above observations there are more c's than a's. So a knight move path of length 16 squares visiting all squares exactly once is not possible. This is the answer of the first question.

The second question is about a knight move path from 1 to 15. This visits all cells except for one. Due to the reasoning above, the unvisited cell should be an a, so the entire path then consists of 3 a's, 4 c's and 8 b's. Using the reasoning above, this is only possible if there are two groups of 4 b's each, separated by a single c or the pattern cac.

Now we consider the additional requirement that the two leftmost numbers on the top row add up to 31. This is only possible if they are 15 and 16. As we saw above the remaining cell left after the knight move path of

1 through 15 is of type a, so a corner cell. So the number 16 will be on the top left. The b cell next to that will contain 15, and is the last cell of the knight move path from 1 to 15. So the sequence ends in a b. It is easily figured out that all the collected requirements only hold for the sequence *acacbbbbcacbbbb*. This only gives a limited amount of remaining options, to be figured out by hand. One possible solution (there are several) is the following:

16	15	8	3
7	4	11	14
12	9	2	5
1	6	13	10

Challenge: finiteness in the spiral sequence

In Chapter 8, we considered the spiral sequence

$$\texttt{spir} = \langle n \rangle = 11010010^3 10^4 10^5 10^6 1 \cdots$$

and we claimed that its turtle figure for the anlges $h(0) = 5$ and $h(1) = 108$ is finite. The full turtle figure was already given there; here we give the turtle figure of the initial part of spir of length 299,300 for the same angles:

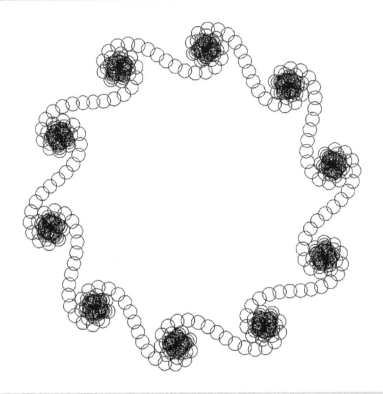

Challenge:

At a first glance, this looks equal to the full turtle figure as given earlier. However, if you look carefully you may see that it is not exactly the same: this initial part does not yet give the full turtle figure. Therefore the first part of the challenge is: spot the difference. Similar to picture puzzles with the goal *find the ten differences*, only now there is only one difference.

The second and more serious part of the challenge is: show that at the initial part of length 320,000 the complete turtle figure of the infinite sequence `spir` is obtained.

More finite turtle figures

We already saw quite some finite turtle figures for several infinite sequences: for periodic sequences, for ultimately periodic sequences, for the spiral sequence presented in a challenge, and a great number for the Thue-Morse sequence.

Here we recall the key theorems to conclude finiteness:

Theorem 6.1: If $h(u)$ is rational and $h(u) \neq 0$, then the turtle figure of the periodic sequence u^∞ is finite.

This theorem was easily extended to ultimately periodic sequences by Theorem 6.2.

Theorem 9.1: If $h(u) = h(v) = 0$ and $P(u) = P(v) = (0,0)$, and a is a sequence in $\{u,v\}^\infty$, then the turtle figure of a is finite.

A new theorem

We are now going to make a kind of combination of these two key theorems. The result can be seen as a variant of Theorem 9.1 in which the condition $h(v) = 0$ and $P(v) = (0,0)$ remains, but the condition for u is substantially weakened: the only requirement is that $h(u)$ is rational, and not equal to 0.

Let's see how the turtle figure of a looks like if a is a sequence in $\{u,v\}^\infty$. If we forget the occurrences of v in a for the moment, and we assume that there are infinitely many occurrences of u, then we know from Theorem 6.1 that the turtle figure of u^∞ is finite. More precisely: if $h(u) = \frac{360m}{n}$ and $h(u) \neq 0$, then the turtle figure of u^∞ is finite, and equal to that of u^n, and consists of at most kn different line segments, where k is the length of u.

DOI: 10.1201/9781003466000-10

But what happens to the turtle figure of a if we do not forget the occurrences of v in a? Then a copy of the turtle figure of v is drawn for every occurrence of v. But according to $h(v) = 0$ and $P(v) = (0,0)$ after processing any occurrence of v, the turtle is in the same position as before with the same angle. The same turtle figure of u^∞ continues to be drawn. Only at the positions where some v occurs between the u's, a copy of the turtle figure of v is drawn. This only occurs on the n positions after the end of a u because after u^n the position and angle is back at the original situation. If k' is the length of v, then the turtle figure of v consists of at most k' line segments, and at most n copies of the turtle figure of v are drawn. As all ingredients show up to be finite, the total number of segments originating from u is at most kn, and the number of segments originating from v is at most $k'n$, thus proving the following theorem:

Theorem 10.1 *Let u and v be words of lengths $k, k' > 0$ over 0 and 1, and let a be a sequence in $\{u, v\}^\infty$. Furthermore $h(u) = \frac{360m}{n}$ and $h(u) \neq 0$, and $h(v) = 0$ and $P(v) = (0,0)$. Then the turtle figure of a is finite and consists of at most $(k + k')n$ line segments.*

In this chapter, we will see a great number of finite turtle figures, where finiteness follows from this theorem. In particular, we focus on $a = f^\infty(0)$ being a purely morphic sequence, for which we remind Theorem 9.2:

Theorem 9.2: Every purely morphic sequence $a = f^\infty(s)$ over 0 and 1 is in $\{f^n(0), f^n(1)\}$, for every natural number n.

In our examples, the morphism f and the angles $h(0)$ and $h(1)$ are chosen carefully guided by the requirement that the conditions of Theorem 10.1 hold for $u = f^n(0)$ and $v = f^n(1)$ for a suitable value of n.

Two equal consecutive symbols

A simple way to satisfy the conditions of Theorem 10.1 is by choosing $f(1) = 11$ and $h(1) = \frac{360m}{2^n}$, and $v = f^n(1) = 1^{2^n}$. Then $h(v) = h(f^n(1)) = h(1^{2^n}) = 360m$, being equal to 0 as an angle. As in the proof of Theorem 6.1 we obtain $P(v) = (0,0)$. The remaining requirement in order to apply Theorem 10.1 to $a = f^\infty(0)$ is that $h(u)$ should be rational and unequal to 0. To complete the definition of a, how to choose $f(0)$? The word $f(0)$ should start in 0 for $a = f^\infty(0)$ to be well defined, and it should contain a 1 in order to force that a is not the sequence 0^∞. In case $f(0) = 01^k$ then a becomes the sequence 01^∞ that we want to avoid, so apart from the first symbol 0 in $f(0)$ there should be at least one more 0. So let's choose $f(0) = 010$. If we now choose $h(0) = 50$ and $h(1) = \frac{-7 \times 360}{2^4} = -157\frac{1}{2}$, then all conditions are met and applying Theorem 10.1 on $a = f^\infty(0)$ with $u = f^4(0)$ and $v = f^4(1) = 1^{16}$ states that the resulting turtle figure is finite. It looks like this:

$$f(0) = 010, f(1) = 11, h(0) = 50, h(1) = -157\frac{1}{2}$$

Here we have

$$u = f^4(0) = 01011010111010110101111111101011010111101011010.$$

Since u consists of 16 zeros and 32 ones, and $h(1^{32}) = 32 \times \frac{-7 \times 360}{16} = -14 \times 360 = 0$ we get $h(u) = h(0^{16}) = 16 \times 50 = 800 = 2 \times 360 + 80 = 80$. Hence the figure has a 9-fold symmetry, because $80 = \frac{2 \times 360}{9}$. In this 9-fold symmetry we see nine 16-point stars inside, caused by the occurrences of $v = f^4(1) = 1^{16}$. And on the outer edge, we see nine 8-point stars that are not quite regular, which are half 16-point stars caused by the 1^8 as it appears in the center of u.

According to Theorem 10.1, the number of line segments of this turtle figure is at most $(k + k')n$. Here we have $n = 9$, $k = 48$ and $k' = 16$, so that yields the bound $(48 + 16)9 = 576$. Contrary to examples of previous theorems, this does not mean that the turtle figure is equal to the turtle figure of the first 576 symbols of a because the number of occurrences of u and v is not balanced. In this example, it turns out that the first 768 symbols are needed for the full turtle figure.

We will give some more finite turtle figures of this type. In fact, we have already done that: in the introductory chapter, we saw a turtle figure of a morphic sequence $f^\infty(0)$ with 17-fold symmetry: it was obtained by again choosing $f(0) = 010$ and $f(1) = 11$, and choosing the angles $h(0) = \frac{6 \times 360}{17}$ and $h(1) = \frac{3 \times 360}{64}$.

It is real fun to play with this: just choose a morphism f and angles such that the requirements of the theorem hold, and generate the corresponding turtle figure, every time resulting in different figures. We will see several examples of this in the following pages, starting with $h(0) = 125$ and $f(0) = 010$ and $f(1) = 11$ and $h(1) = \frac{3 \times 360}{64} = 16\frac{7}{8}$.

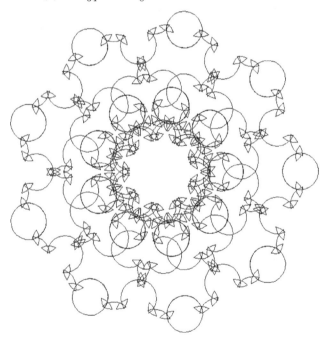

$$f(0) = 010, f(1) = 11, h(0) = 125, h(1) = 16\frac{7}{8}$$

Choosing $f(0) = 010$ was just the first and simplest choice. Of course this may be changed, for instance by $f(0) = 0101$.

$$f(0) = 0101, f(1) = 11, h(0) = -132, h(1) = 33\frac{3}{4}$$

Next we change $f(1)$ and $h(1)$: choosing $f(1) = 111$ yields $f^n(1) = 1^{3^n}$, and by choosing $h(1)$ to be a fraction having a power of three in the denominator Theorem 10.1 will be applicable in a similar way. One example with $h(1) = \frac{10 \times 360}{27} = 133\frac{1}{3}$ and having a 27-fold symmetry is the following:

$$f(0) = 010, f(1) = 111, h(0) = -120, h(1) = 133\frac{1}{3}$$

When considering all these finite turtle figures, this is a perfect moment for investigating the challenge on finiteness of a turtle figure of the spiral sequence

$$\text{spir} = \langle n \rangle = 110100 10^3 10^4 10^5 10^6 1 \cdots .$$

Clearly spir is not periodic, but in the challenge we have $h(0) = 5$, so $h(0^{72}) = 360$. That means that the turtle figure of 0^{72} is a regular 72-sided polygon, nearly a circle to the eye, and after every 0^{72} the turtle is back where it started, both in position and angle. The crucial observation now is that spir is obtained from the periodic sequence u^∞ for u defined by

$$u = 110100 10^3 0^4 10^5 1 \cdots 10^{71}$$

by plugging in 0^{72} More precisely, in the second occurrence of u this is done once after every 1, in the third occurrence of u this is done twice after every

1, and so on. To determine the size of the turtle figure of u^∞ we need to determine $h(u)$. The word u consists of 72 ones and $1 + 2 + 3 + \cdots + 71$ zeros. So the number of zeros in u is a triangular number. In Chapter 3 we derived a formula for that: $1 + 2 + 3 + \cdots + 71 = \frac{71 \times 72}{2}$. Of course this may be calculated, but we don't need to because $h(0^{72}) = 360$, which is equal to 0 as an angle. That means that $h(0^{\frac{k \times 72}{2}})$ equals 0 if k is even, and equals 180 if k is odd. So for $k = 71$ it is $180°$. We had $h(1) = 108 = \frac{3 \times 360}{10}$, so

$$h(1^{72}) = \frac{72 \times 3 \times 360}{10} = \frac{108 \times 360}{5} = \frac{3 \times 360}{5} = 216,$$

where we subtracted $360°$ from the angle a number of times in the penultimate step. So $h(u) = 180 + 216 = 396 = 36 = \frac{360}{10}$. According to the basic theorem about periodic sequences from Chapter 7, Theorem 6.1, the turtle figure of u^∞ is equal to that of u^{10}, and consists of at most $10k$ line segments where k is the length of u. This number 10 is also reflected in the 10-fold symmetry of the presented turtle figure. The turtle figure of `spir` is obtained from the turtle figure of u^∞ by adding extra regular 72-sided polygons on several places. The first time u is run through, this does not yet happen. From that first time on it happens every time, a growing number of times, but always at the same points. Whether at such a point the regular 72-sided polygon is drawn once, twice or a hundred times, does not matter for the resulting turtle figure. So after going through u 10 more times after the first time, and draw a regular 72-sided polygon at all those extra places, we have the full turtle figure of `spir`. The question was which initial part of `spir` is needed before completion. The just sketched part consists of 11 times u, with the groups of 72 zeros added. The easiest way to determine that length is to observe that after 11 times u we have exactly $11 \times 72 = 792$ ones. The number of zeros from the initial part up to and including the 792nd one is

$$0 + 1 + 2 + 3 + \cdots + 791 = \frac{791 \times 792}{2} = 313,236,$$

where we again conveniently use the formula for triangular numbers. Together with the 792 ones this gives a number clearly being less than 320,000, solving the second part of the challenge. To observe that more than 299,300 are needed, please carefully look at the circles on the far right of the corresponding picture. To be precise: at the top of the two rightmost bumps, one circle is not completely finished at the bottom right of that bump. This observation solves the first part of the challenge.

One more challenge about spiral sequences is still open: is it possible to present

$$\langle n^2 \rangle = 11010^4 10^9 10^{16} 1 \cdots$$

as a morphic sequence? Indeed this will be possible. We already know that $\mathtt{spir} = \langle n \rangle$ is a morphic sequence: we have $\mathtt{spir} = c(f^\infty(2))$ for

$$f(0) = 0, \ f(1) = 01, \ f(2) = 21, \ c(0) = 0, \ c(1) = c(2) = 1.$$

A first step is just slightly modify this definition and see what happens. For instance, defining $f(1) = 00$ and leaving everything else the same, gives the sequence $\langle 2n \rangle$. Similar modifications may yield $\langle 5n + 3 \rangle$, but trying to make $\langle n^2 \rangle$ using only these three symbols 0, 1 and 2 appears to fail. More power is obtained by adding an extra symbol 3 that eventually will be mapped to 0 by the coding c. This can be done in several ways. One possible way is $c(f^\infty(2))$ for

$$f(0) = 0, \ f(1) = 31, \ f(2) = 21, \ f(3) = 003,$$

$$c(0) = c(3) = 0, \ c(1) = c(2) = 1.$$

Indeed this gives

$$f^\infty(2) = 2131003310030030331 \cdots$$

which looks like $\langle n^2 \rangle$ after applying c, so replacing every 2 by 1, and every 3 by 0. But how do we know for sure? We obtain

$$f^\infty(2) = 21f(1)f^2(1)f^3(1)f^4(1) \cdots,$$

where each $f^k(1)$ is of the shape $v_k 1$, where v_k consists only of 0 and 3. Writing $|v|_s$ for the number of times the symbol s occurs in a word v, we have to show that $|v_k|_0 + |v_k|_3 = k^2$ for every k. We now show that for every k we have $|v_k|_0 = k^2 - k$ and $|v_k|_3 = k$. Because of $f(1) = 31$ we have $|v_1|_0 = 0 = 0^2 - 0$ and $|v_1|_3 = 1$, so for $k = 1$ the statement is correct. And from $f(0) = 0$, $f(1) = 31$ and $f(3) = 003$ and $f(v_k 1) = v_{k+1} 1$ follows $|v_{k+1}|_0 = 2|v_k|_3 + |v_k|_0$ and $|v_{k+1}|_3 = 1 + |v_k|_3$. Using these properties it is straightforward to prove by induction that $|v_k|_0 = k^2 - k$ and $|v_k|_3 = k$ for every k, similar to the proof of the formula for triangular numbers we gave long ago. So $|v_k|_0 + |v_k|_3 = k^2$ for every k, hence indeed $c(f^\infty(2)) = \langle n^2 \rangle$, solving the challenge.

In the last example, we gave before this excursion to spiral sequence challenges, we had $f(0) = 010$, $f(1) = 111$, and a power of 3 in the denominator of $h(1)$, by which Theorem 10.1 could be applied. However, also variants with finite turtle figures show up where this theorem does not apply. For instance, choosing $f(0) = 010$, $f(1) = 111$, $h(0) = -120$ and $h(1) = \frac{17 \times 360}{54}$ yields the following turtle figure for $f^\infty(0)$:

The reason that this turtle figure is finite is related to the argument for the spiral sequence. Observe that $h(f^3(1)) = h(1^{27}) = 180$. This is not zero, but twice this angle is zero. The sequence $f^\infty(0)$ may be obtained from the periodic sequence

$$(0101^3 0101^9 0101^3 0101^{27})^\infty$$

by adding 1^{54} a number of times to some occurrences of 1^{27}. Since $h(1^{54}) = 0$ and $P(1^{54}) = (0,0)$, the resulting turtle figure of $f^\infty(0)$ is finite.

Rosettes

Every year an international conference called *Bridges* is organized about the connection between art and mathematics. In 2014 a contribution by Paul Gailiunas entitled *Recursive Rosettes* was essentially about a special class of turtle figures. More precisely, in our notation it was about finite turtle figures of the sequence

$$f^\infty(0) = 011000110110110001100\cdots$$

for the morphism f defined by $f(0) = 011$ and $f(1) = 0$. These are called *rosettes*. It is surprising to see that by choosing very simple angles just being integer numbers of degrees, very nice finite symmetrical figures show up. We give an example that was also given by Gailiunas in his article: $h(0) = 140$ and $h(1) = -40$ gives the following turtle figure

$$f(0) = 011, f(1) = 0, h(0) = 140, h(1) = -40$$

We will now show that the finiteness of this turtle figure follows from Theorem 10.1. However, this is somewhat more complicated than our first applications. Since in $f(1)$ there is no 1 and in $f(0)$ the symbol 1 only appears in pairs, we conclude that $f^\infty(0)$ is composed from 0 and 11, so is in $\{0, 11\}^\infty$. And since for every k we have $f^\infty(0) = f^k(f^\infty(0))$, the sequence $f^\infty(0)$ is also in $\{f^k(0), f^k(11)\}^\infty$ for any k. We will now apply Theorem 10.1 to $u = f^k(0)$ and $v = f^k(11) = f^k(1)f^k(1)$ for a number k still to be chosen. The rationality requirement of $h(u)$ is easy to satisfy. The remaining requirement is that $h(v) = 0$ and $P(v) = (0,0)$. Since the first half and the second half of v are both equal to $f^k(1)$, it suffices to show that $h(f^k(1)) = 180$, an argument that we have seen before.

We want to calculate $h(f^k(1))$. Recall that $|u|_0$ is the number of zeros in a word u, and $|u|_1$ the number of ones in u. Since $f(1)$ contains no 1 and $f(0)$ contains two ones, we have $|f^{k+1}(1)|_1 = 2|f^k(1)|_0$. Since $f(1)$ and $f(0)$ each contain one 0, we have $|f^{k+1}(1)|_0 = |f^k(1)|_0 + |f^k(1)|_1$. This applies to every k. Since $f^1(1) = f(1) = 0$ we have $|f^1(1)|_0 = 1$ and $|f^1(1)|_1 = 0$. These properties easily yield the following table:

k	$|f^k(1)|_0$	$|f^k(1)|_0$
1	1	0
2	1	2
3	3	2
4	5	6

k	$|f^k(1)|_0$	$|f^k(1)|_0$
5	11	10
6	21	22
7	43	42
8	85	86
9	171	170
10	341	342

In particular we obtain $|f^8(1)|_0 = 85$ and $|f^8(1)|_1 = 86$. So for $h(0) = 140 = \frac{7 \times 180}{9}$ and $h(1) = -40 = \frac{-2 \times 180}{9}$ this yields

$$h(f^8(1)) = \frac{(85 \times 7 - 86 \times 2) \times 180}{9} = \frac{423 \times 180}{9} = 47 \times 180,$$

indeed an odd number times $180°$, so equal to $180°$. After checking that $h(f^8(0)) \neq 0$, by Theorem 10.1 we conclude that this turtle figure is finite.

In this example we ended up with integer numbers of degrees because we had nine as the denominator. But the same method also works for other numbers in the denominator. In the following example we do not give the complete turtle figure, but focus on a part of it in order to see the details properly. As expected, the complete turtle figure is a symmetrical figure similar to the other examples in this chapter:

$$f(0) = 011, f(1) = 0, h(0) = \frac{7 \times 180}{13}, h(1) = \frac{-2 \times 180}{13}$$

More symbols

We have already saw turtle figures containing 16-point stars caused by $f(1) = 11$. Similarly, one makes turtle figures with regular octagons or 16-sided polygons. Figures containing 9-point stars, or 27-point stars, may be obtained by choosing $f(1) = 111$. Is it possible to make combinations of these, for example both containing regular octagons and 9-point stars? Using only two symbols 0 and 1, that will be tricky: we need the 0 to start up, and then $f(0)$ will need to contain at least a 0 and at least a 1. And for $f(1)$ we can't choose 11 and 111 at the same time. But the same ideas apply for morphic sequences over more than two symbols, say, 0, 1, and 2. In order to conclude finiteness for this case, we will extend Theorem 10.1. Then we again use 0 to start up, and hope that choosing $f(1) = 11$ and $h(1) = 45$ will yield several regular octagons, and $f(2) = 222$ and $h(2) = 160$ will yield several 9-point stars.

Assume we have k symbols, numbered from 0 to $k - 1$. Then $f^\infty(0)$ is in $\{f^n(0), f^n(1), \ldots, f^n(k - 1)\}^\infty$, again for any n to be chosen. Here $\{u_0, u_1, \ldots, u_{k-1}\}^\infty$ is the set of sequences composed from words from the set $\{u_0, u_1, \ldots, u_{k-1}\}$ in any order. So this extends the notation $\{u, v\}^\infty$ we saw earlier. This notation gives the following more general version of Theorem 10.1:

Theorem 10.2 *Let $k \geq 2$, and let k words $u_0, u_1, \ldots, u_{k-1}$ be given. Let a be a sequence in $\{u_0, u_1, \ldots, u_{k-1}\}^\infty$. Assume that $h(u_0)$ is rational and $h(u_0) \neq 0$, and for every $i = 1, \ldots, k - 1$ we have*

- *u_i occurs only finitely often in a, or*
- *$h(u_i) = 0$ and $P(u_i) = (0,0)$.*

Then the turtle figure of a is finite.

Proof: First we consider the u_i that occur only finitely often in a. Because they only occur finitely often in a, all those finitely many occurrences are in some finite initial part of a. The turtle figure of a now consists of the finite turtle figure of this finite initial part, to which the turtle figure of the rest of a will be added. So we only have to show that the turtle figure of that remaining part of a is finite. This remaining part consists of u_0 and zero or more u_i for which $h(u_i) = 0$ and $P(u_i) = (0,0)$. Just like in the proof of Theorem 10.1, the remaining turtle figure consists of the turtle figure of u_0^∞ which is finite according to Theorem 6.1, to which a finite number of copies of finite turtle figures of u_i is added. So the full turtle figure of a consists of a combination of finitely many finite parts, so is finite itself. □

The first few attempts to make a turtle figure of a morphic sequence containing both regular octagons and 9-point stars according to this theorem failed, as the stars and octagons mixed up and could not be distinguished from each other. But by giving one of the angles a minus sign it works. More precisely, by choosing $f(0) = 0120$, $f(1) = 11$ and $f(2) = 222$, and $h(0) = 30$, $h(1) = 45$ and $h(2) = -160$ the conditions of Theorem 10.2 hold, and gives the following finite turtle figure for $f^\infty(0)$:

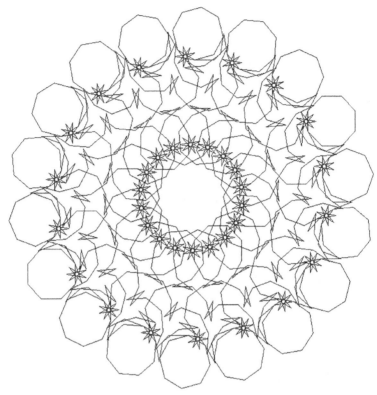

$$f(0) = 0120, f(1) = 11, f(2) = 222, h(0) = 30, h(1) = 45, h(2) = -160$$

Indeed, two groups of 18 9-point stars are clearly recognized, and 18 regular octagons at the outer edge. This number 18, and the 18-fold symmetry is caused by the fact that $18h(f^3(0))$ is a multiple of $360°$.

We give another application of Theorem 10.2, this time a figure with 22-fold symmetry, by choosing $f(0) = 0021$, $f(1) = 11$ and $f(2) = 22$, $h(0) = \frac{3 \times 360}{11}$, $h(1) = \frac{7 \times 360}{16}$ and $h(2) = \frac{360}{32}$. Here $h(1)$ creates a great number of 16-point stars, and $h(2)$ creates several regular 32-sided polygons, which look like circles:

$$f(0) = 0021, f(1) = 11, f(2) = 22, h(0) = \frac{3 \times 360}{11}, h(1) = \frac{7 \times 360}{16}, h(2) = \frac{360}{32}$$

Adding a tail

Theorem 10.2 allowed another case that we did not yet exploit: if u_i occurs only finitely many times in the sequence, then for u_i there are no requirement on the angles. Here we will give an example using this.

In the first example of this chapter, we had $f(0) = 010, f(1) = 11, h(0) = 50$ and $h(1) = -157\frac{1}{2} = \frac{-7 \times 360}{2^4}$. If we change this to $f(0) = 012$, $f(1) = 11$, $f(2) = 212$, $h(0) = 50$, $h(1) = -157\frac{1}{2} = \frac{-7 \times 360}{2^4}$ and $h(2) = 50$, then we get exactly the same turtle figure, because 0 and 2 behave exactly the same. But now we see that in the sequence $f^\infty(0)$ the symbol 0 only appears at the first position, and nowhere else. And for every k, $f^\infty(0)$ is composed from the words $f^k(0)$, $f^k(1)$, and $f^k(2)$. But from this we see that $f^k(0)$ only occurs finitely often in $f^\infty(0)$, namely only at the beginning. According to Theorem 10.2 we have no requirements on the angle of $f^k(0)$. So when changing the definition of $f(0)$, Theorem 10.2 still applies, as long as $f^k(0)$ only occurs at the beginning of $f^\infty(0)$. That is the case in the following modification where we choose

$$f(0) = 02122, \ f(1) = 11, \ f(2) = 212,$$
$$h(0) = 50, \ h(1) = -157\frac{1}{2} = \frac{-7 \times 360}{2^4}, \ h(2) = 50.$$

This results in the following turtle figure:

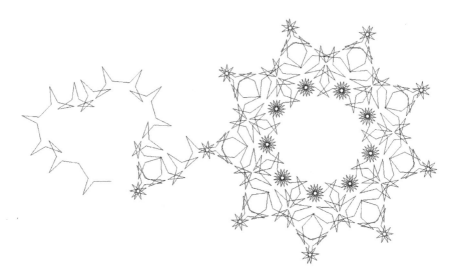

Indeed, we see the same turtle figure as before, but now an irregular tail has been added in the front because of $f(0) = 02122$. On the one hand, it is an anticlimax: after all beautiful symmetrical figures in this chapter, we end up with such a mess. On the other hand, it is good to know that this is possible, and we have a theorem yielding non-symmetric finite turtle figures for morphic sequences. It has some similarity with the symmetric finite turtle figures for periodic sequences that were extended to non-symmetric turtle figures for ulitmately periodic sequences.

Challenge: one more finite turtle figure

The turtle figure for $f^\infty(0)$ with $f(0) = 01$ and $f(1) = 00$, and angles $h(0) = 140$ and $h(1) = -80$ is as follows:

That's quite surprising for such a very simple definition of *f* and such simple angles: it looks finite and symmetrical, just like the several figures we saw in this chapter.

Challenge:

Show that the turtle figure for $f^\infty(0)$ is indeed finite, and consists of <10,000 line segments.

Our main tool for showing that such a rosette-like turtle figure is finite is Theorem 10.1, so an obvious approach would be to look for words *u* and *v* to which that theorem applies. Good luck!

Fractal turtle figures

Although the sequences we consider are infinite, until now the focus was on *finite* turtle figures of such sequences. In this chapter, we switch to particular infinite turtle figures, having a lot of structure. Because they are infinite, only finite initial parts will be shown, but these finite parts will show surprising patterns, being very different from the patterns we have seen so far.

The emphasis will be on turtle figures having the remarkable property that the result of magnifying the figure by a fixed factor will be contained in the original figure. This phenomenon is closely related to figures called *fractals*, so the resulting turtle figures will be called *fractal*. But let's start with some general remarks on fractals.

Mandelbrot sets

A well-known example of fractals are the images of the *Mandelbrot set*, named after the French-Polish mathematician Benoît Mandelbrot (1924–2010) who studied their fractal patterns. Apart from the fractal behavior, they have no further relationship with turtle figures. The Mandelbrot set is a set of points in the plane that is described as a set of *complex numbers*. In an earlier chapter on numbers, we already discussed complex numbers: numbers written as $x + iy$ where $i^2 = -1$, and x, y are real numbers. Any complex number c defines the sequence $c_0, c_1, c_2, \ldots$ of complex numbers as follows:

$$c_0 = 0, \quad \text{and} \quad c_{n+1} = c_n^2 + c \text{ for all } n \geq 0.$$

Here the calculation $c_n^2 + c$ is done in the complex numbers. If you are not used to that, you may also write $c = (x, y)$ and $c_n = (x_n, y_n)$, and define

$$x_0 = 0, \ y_0 = 0, \quad \text{and}$$

DOI: 10.1201/9781003466000-11

$$x_{n+1} = x_n^2 - y_n^2 + x, \ y_{n+1} = 2x_ny_n + y \text{ for all } n \geq 0.$$

Identifying (x, y) with $x + iy$ shows that this coincides with the definition in complex numbers.

Now for a complex number c the corresponding sequence $c_0, c_1, c_2, \ldots$ may be *bounded*, that is, there exists a number N such that $x_n^2 + y_n^2 < N$ for each $n \geq 0$. Otherwise, the sequence is *unbounded*. The Mandelbrot set is defined to be the set of complex numbers c for which the corresponding sequence $c_0, c_1, c_2, \ldots$ is bounded.

Of course one may define whatever he or she wants, but this particular set is quite complicated and has some remarkable properties. Let's try to do and attempt to determine which complex numbers c, being interpreted as points in the plane, are in this Mandelbrot set and which are not. To do this exactly is not so easy: every finite initial part of the sequence $c_0, c_1, c_2, \ldots$ is bounded, so by only considering any finite part of the sequence it is impossible to determine whether the sequence is bounded or not. In the definition we do not know the size of N: if the sequence sways up and down a lot at the beginning this may suggest that it is not bounded, but even then it can be bounded for a very large value for N. But we may try to approximate. To do so, we start by computing the initial part $c_0, c_1, c_2, \ldots, c_{99}$. If all of these 100 values satisfy $x_n^2 + y_n^2 < 10$, then we guess that this c is in the set, otherwise not. Calculating this guessed approximation of the Mandelbrot set by a simple Lazarus program gives the figure shown on the next page.

Here the values of x in $c = x + iy$ run from $\frac{-3}{2}$ to $\frac{1}{2}$, from left to right in this picture. The values of y run from $\frac{-3}{2}$ to $\frac{3}{2}$, from bottom to top in this image. The points we expect to be in the Mandelbrot set, so with $x_n^2 + y_n^2 < 10$ for n to 100, are the white points in the inner region. We expect the black points not to be in the Mandelbrot set: there the boundary $N = 10$ is exceeded in the first 100 elements of the sequence $c_0, c_1, c_2, \ldots$ In the white outer area, the $N = 10$ is already exceeded in the first ten elements of the sequence $c_0, c_1, c_2, \ldots$, so we know for sure that those points are not in the Mandelbrot set. Although this is only a very rough approximation, typical features of this Mandelbrot set may be observed: on the right we see a heart-shaped figure, surrounded by all kinds of circles in different sizes, while these circles are touched by smaller circles. And now the remarkable observation comes in: zooming in on these smaller circles, with a less rough approximation and a finer scale, a similar pattern of circles and smaller touched circles shows up. And this property that zooming in or out yields a figure being similar to the original one, is the key property of fractals.

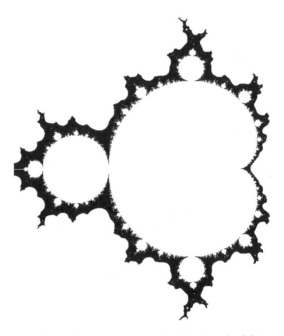

Calculating the Mandelbrot more precisely instead of this rough approximation requires a lot of computation time. The theory explaining the above sketched fractal behavior is quite tricky. In fact, the Mandebrot set does not even have exact fractal behavior, in the sense that zooming in yields exactly the same set. It turns out that the result of zooming in is similar, but not exactly equal. When zooming in further, not only the same patterns reappear, but sometimes also new patterns. In a way, this makes it even more exciting. But for us the only goal was to give a flavor of fractal behavior, so for more information on the Mandelbrot set the reader should consider other sources. We will switch to turtle figures now.

Fractal turtle figures

Before giving examples of fractal turtle figures, first we precisely define when a turtle figure will be called fractal. As turtle figures consist of line segments all having the same length, it is more natural to focus on similarity after zooming out, than after zooming in. But for our definition, we want to allow both: zooming in is multiplying by a factor c satisfying $0 < c < 1$, and zooming out is multiplying by a factor c satisfying $c > 1$. As a first attempt a set of points in the plane is called *fractal* if there exists a positive real number $c \neq 1$ such that for every point (x, y) in that set, the point $c(x, y) = (cx, cy)$ is also in that set. This definition is very general: for example, a line or the full plane is also fractal. But this definition also has a limitation: it only considers zooming out by a factor c, without rotation. It is reasonable that if zooming out is combined by a rotation, to call the result fractal too. Later we will see examples of this. In Chapter 10, we introduced the notation R_h for rotating over an angle h. Again using this notation, we arrive at the following definition of a fractal set:

Definition 11.1 A set of points in the plane is called *fractal* if there exists a positive real number $c \neq 1$ and an angle h such that for each point (x, y) in that set, the point $cR_h(x, y)$ is also in that set.

An important observation is that the order of rotating over an angle h and scaling by a factor of c does not matter, since we always have

$$cR_h(x, y) = R_h(c(x, y)) = R_h(cx, cy).$$

This definition is still very general, including lines and the full plane, but it captures exactly the core property that we want for turtle figures. So this is the definition of a fractal set as we will use it.

The number $c \neq 1$ is called the *scaling factor*, and the angle h is called the *rotation angle* of the fractal set. The definition of fractal turtle figure is directly derived from this:

Definition 11.2 A turtle figure of a sequence is called *fractal* if the set of end points of the line segments is a fractal set.

A first and very important observation is that every fractal set containing at least one point $(x, y) \neq (0, 0)$ is always *infinitely large*. Indeed, according to the definition of a fractal set, all points

$$(x, y), \ cR_h(x, y), \ c^2R_{2h}(x, y), \ c^3R_{3h}(x, y), \ldots$$

are also in that set. These points are all distinct since they all have different distances to $(0, 0)$ due to $c \neq 1$. For example, if $c = 2$ and $h = 0$, then $R_h(x, y) = (x, y)$ for every x, y, and $(1, 0)$ is in the fractal set, then the infinitely many points

$$(1, 0), \ (2, 0), \ (4, 0), \ (8, 0), \ldots, (2^k, 0), \ldots$$

are all in that same fractal set.

We will now find criteria on a morphism f and angles $h(0)$ and $h(1)$ such that if these criteria hold, then the turtle figure of $f^\infty(0)$ is fractal. Before presenting this in a theorem, first we give an illustration in order to get some feeling on what is going on.

We choose $h(0) = 0$ and $h(1) = 90$, and $f(0) = 001111$ and $f(1) = 10$. How does the turtle figure of the word $f(0) = 001111$ look like? By the first two symbols 00, a line segment of length 2 is drawn. Next, by the four symbols 1111, a square is drawn. After drawing this square, the angle of the turtle is equal to its initial value, since $4h(1) = 360 = 0$. So the turtle figure of $f(0) = 001111$ looks like this:

Here turtle drawing starts in the point $(0, 0)$, being the leftmost point. The original angle is 0, describing the direction pointing to the right. After processing 001111, the turtle has arrived in $(2, 0)$ at the right, again having angle 0 describing the direction pointing to the right.

By processing $f(1) = 10$, first the angle rotates by $90°$, after which a line segment of length 2 is drawn. This makes the turtle figure of $f^2(0) = 00111100111110101010$ as follows:

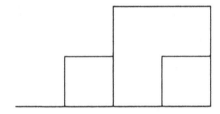

This may be seen as twice the turtle figure of $f(0) = 001111$, completed by a larger square drawn by 0101010, after which the point $(4, 0)$ is reached on the right, with the direction pointing to the right again. Next we consider the turtle figure of $f^3(0)$, which looks like this, in adjusted scale.

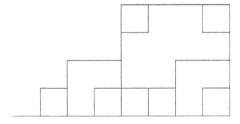

Apart from drawing the squares, the turtle figures of $f(0)$ and $f(1)$ are the same as the turtle figures of 0 and 1, with one crucial difference: the drawn segment has length 2 in both cases instead of length 1. And every next application of f produces a copy of the previous one, magnified by a factor of 2. We already observed this in the given pictures until $f^3(0)$. This behavior will cause that the turtle figure of $f^\infty(0)$ fractal, having scaling factor $c = 2$ and rotation angle $h = 0$. The fractal turtle figure of the sequence $f^\infty(0)$ is infinitely large: it starts on the left, but continues indefinitely to the right with increasingly larger squares. Therefore, we only give a finite initial part, in a somewhat finer scale than the smaller initial parts we already showed.

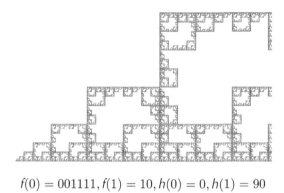

$$f(0) = 001111, f(1) = 10, h(0) = 0, h(1) = 90$$

The main theorem

We are now ready to state and prove a theorem by which indeed can be proved that this example is a fractal turtle figure, having scaling factor 2 and rotation angle 0. This theorem will form the basis of all fractal turtle figures we will see later.

Theorem 11.1 *Let $f(0) = 0u$ and $f(1) = 1u'$ for non-empty words u, u' over 0 and 1, satisfying $h(u) = h(u') = 0$ and $P(u) = P(u') = (x, y)$. Let $c = \sqrt{(x+1)^2 + y^2} \neq 1$. Then the turtle figure of $f^\infty(0)$ is fractal, having scaling factor c and rotation angle h determined by $cR_h(1, 0) = (x + 1, y)$.*

For readers familiar with arctan, the *arctangent*: if $x > -1$ in Theorem 11.1, then $h = \arctan \frac{y}{x+1}$.

In our example, we had $h(0) = 0$ and $h(1) = 90$, and $f(0) = 001111$ and $f(1) = 10$. This satisfies the requirements of Theorem 11.1: for $u = 01111$ and $u' = 0$ it indeed holds that $h(u) = h(u') = 0$ and $P(u) = P(u') = (x, y) = (1, 0)$. Here the scaling factor is $c = \sqrt{(x+1)^2 + y^2} = \sqrt{(1+1)^2 + 0^2} = 2$ and the rotation angle h is 0. Because $c = 2 \neq 1$ we conclude by

Theorem 11.1 that the turtle figure of $f^\infty(0)$ is fractal, having scaling factor $c = 2$ and rotation angle $h = 0$.

Now we give the proof of Theorem 11.1.

Proof: The end points of the segments of the turtle curve of a sequence a are the positions $P(v)$ for the finite initial parts v of a. According to the definition of a fractal turtle curve, we have to show that for every such finite initial part v of $a = f^\infty(0)$ there is an initial part v' of a for which $P(v') = cR_h(P(v))$. We will show that this holds for $v' = f(v)$. Since $a = f(a)$, the word $f(v)$ is also an initial part of a.

We recall Lemma 6.1, stating that $P(uv) = P(u) + R_{h(u)}(P(v))$ for all words u, v. This property we will use a number of times.

We obtain $h(f(0)) = h(0u) = h(0) + h(u) = h(0)$ and $h(f(1)) = h(1u') = h(1) + h(u') = h(1)$.

Since we always start by the direction pointing to the right, so from $(0,0)$ to $(1,0)$, we have $P(0) = R_{h(0)}(1,0)$. Now we obtain

$$
\begin{aligned}
P(f(0)) &= P(0u) = P(0) + R_{h(0)}(P(u)) = R_{h(0)}(1,0) + R_{h(0)}(P(u)) \\
&= R_{h(0)}((1,0) + (x,y)) = R_{h(0)}(cR_h(1,0)) = cR_h(P(0)).
\end{aligned}
$$

Similarly we have $P(1) = R_{h(1)}(1,0)$, and

$$
\begin{aligned}
P(f(1)) &= P(1u') = P(1) + R_{h(1)}(P(u')) = R_{h(1)}(1,0) + R_{h(1)}(P(u')) \\
&= R_{h(1)}((1,0) + (x,y)) = R_{h(1)}(cR_h(1,0)) = cR_h(P(1)).
\end{aligned}
$$

So for both $s = 0$ and $s = 1$ we obtain $P(f(s)) = cR_h(P(s)$ and $h(f(s)) = h(s)$.

Now we show by induction on the length of v that for each word v we have $P(f(v)) = cR_h(P(v))$ and $h(f(v)) = h(v)$. For length 1 we have just shown this. For the induction step we write $v = v's$ then according to the induction hypothesis the property holds for v', yielding

$$
\begin{aligned}
P(f(v)) &= P(f(v')f(s)) = P(f(v')) + R_{h(f(v'))}P(f(s)) \\
&= cR_h(P(v')) + R_{h(f(v'))}(cR_h(P(s))) \\
&= cR_h(P(v') + R_{h(f(v)'))}(P(s))) \\
&= cR_h(P(v') + R_{h(v')}(P(s))) = cR_h(v's) = cR_h(v).
\end{aligned}
$$

This proves the property for every word v, in particular for every finite initial part of a, and that's what we had to prove. □

Such a proof is quite technical and not very simple. But if you really want to be sure that the statement is true, checking such a proof completely is the only way to go. From that moment on, you may trust the theorem and may use it again and again without worrying about these technical steps any more. Theorem 11.1 gives a simple criterion for a turtle figure for being fractal.

Now we will exploit this the other way around: for what kind of examples do the conditions of Theorem 11.1 hold? What would the resulting fractal turtle figure look like?

We have already seen one example, but it is easy to vary. For example, if we choose $h(0) = 0$ and $h(1) = 120$, and $f(0) = 00111$ and $f(1) = 10$, the conditions of Theorem 11.1 are fulfilled in a similar way, but we get triangles instead of squares:

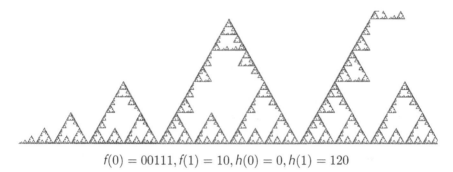

$$f(0) = 00111, f(1) = 10, h(0) = 0, h(1) = 120$$

Similarly, pentagons are obtained by choosing $h(0) = 0$, $h(1) = 72$, $f(0) = 0011111$ and $f(1) = 10$:

$$f(0) = 0011111, f(1) = 10, h(0) = 0, h(1) = 72$$

By doubling $h(1)$ and leaving the rest unchanged, we obtain five-point stars:

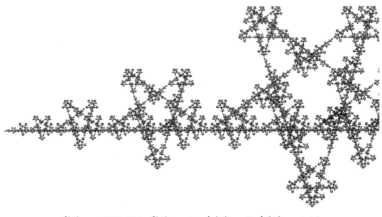

$$f(0) = 0011111, f(1) = 10, h(0) = 0, h(1) = 144$$

All these examples are fractal according to Theorem 11.1, having rotation angle 0 and scaling factor 2. They start on the left and continue infinitely to the right, while we only show only a finite initial part. This allows an endless number of variations, even restricting to rotation angle 0 and scaling factor 2. For example, similarly applying Theorem 11.1 yields the fractal turtle figure of $f^{\infty}(0)$ for $h(0) = 120$, $h(1) = 0$, $f(0) = 00001$ and $f(1) = 11$:

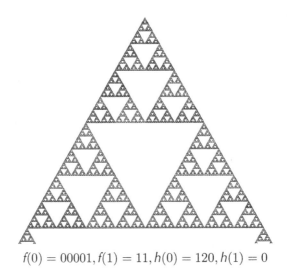

$$f(0) = 00001, f(1) = 11, h(0) = 120, h(1) = 0$$

Here the turtle figure starts at the top and continues downwards indefinitely, of which we only show a finite part. Essentially, this is the *Sierpinski triangle*, discovered and described by the Polish mathematician Wacław Sierpinski (1882–1969).

In the same way, still having rotation angle 0, but now having scaling factor 3, we obtain the fractal turtle figure of $f^\infty(0)$ for $h(0) = 90$, $h(1) = 0$, $f(0) = 0000011$ and $f(1) = 111$. Here the fractal behavior appears nicely: the squares-within-squares that start at the bottom left inflate by a factor of three and repeat.

$$f(0) = 0000011, f(1) = 111, h(0) = 90, h(1) = 0$$

Examples with rotation

Until now we only saw examples having rotation angle 0, but Theorem 11.1 also allows for fractal turtle figures having rotation angle unequal 0. In this section, this will be elaborated. As a first example, we consider the turtle figure of $f^\infty(0)$ for $f(0) = 0101111$, $f(1) = 110$, $h(0) = 90$ and $h(1) = -90$. Here the conditions of Theorem 11.1 hold: for $u = 101111$ and $u' = 10$ we obtain $h(u) = h(u') = 0$. Next we determine $P(u)$ and $P(u')$. Starting with the direction pointing to the right, due to $h(1) = -90$ we first turn a right angle clockwise, by which the direction is down. A unit step then yields $P(1) = (0, -1)$. And from there $h(0) = 90$ causes $P(u') = P(10) = (1, -1)$. Since 1111 draws a square and returns to the same point, we have $P(u) = P(101111) = P(10) = (1, -1)$. The conditions of Theorem 11.1 are now met, and the turtle figure is fractal having scaling factor $c = \sqrt{2^2 + 1^2} = \sqrt{5}$ and rotation angle $h = \arctan \frac{-1}{2} \approx -26,565$.

To understand what is happening, *complex numbers* prove to be surprisingly helpful. As a reminder, a complex number can be written as $x + iy$ where x, y are real numbers, and i is a number satisfying $i^2 = -1$. The point (x, y) in the plane is identified with this complex number $x + iy$. And now the key idea comes in: the combination of rotation over an angle h and scaling up by a factor of c corresponds to multiplying by a fixed complex number. In our example with scaling factor $c = \sqrt{2^2 + 1^2} = \sqrt{5}$ and rotation angle $h = \arctan \frac{-1}{2} \approx -26,565$ this is multiplying by $2 - i$, that is $1+$ the complex number corresponding to $P(u) = (1, -1)$. It is not a goal of this book to present the theory of complex numbers in detail, but we want to show how this works in practice, and how positions of the points $P(f^k(0))$ may be computed by complex multiplication for increasing k. We give the turtle figure of $f^4(0)$, so an initial part of the turtle figure of $f^\infty(0)$. The coordinates of a number of points are indicated in the picture. We start in $(0, 0)$. To determine $P(0)$, we first turn the direction to the right $90°$, counterclockwise, so that becomes up, and do one step, yielding $P(0) = (0, 1)$. As a complex number this is $0 + i = i$. According to the theorem the turtle arrives in $P(f(0)) = P(0101111)$ by rotation and scaling up. This is equivalent to multiplying by $2 - i$. Indeed, multiplying by $2 - i$ yields $i(2 - i) = 1 + 2i$, corresponding to $P(f(0)) = (1, 2)$.

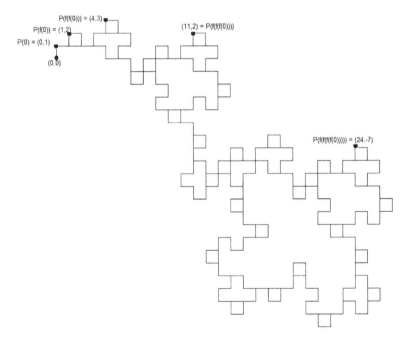

Continuation gives the following:
$(1 + 2i)(2 - i) = 4 + 3i$, so $P(f^2(0)) = (4, 3)$,
$(4 + 3i)(2 - i) = 11 + 2i$, so $P(f^3(0)) = (11, 2)$, and
$(11 + 2i)(2 - i) = 24 - 7i$, so $P(f^4(0)) = (24, -7)$.

In the picture, we see these points indicated. The first part of the fractal behavior is visible: every pattern appears in a modified variant in every next step, being enlarged by a factor $\sqrt{5}$ and slightly rotated clockwise.

This fractal behavior is much more visible in the turtle figure of a much larger initial part of $f^\infty(0)$ in a much finer scale:

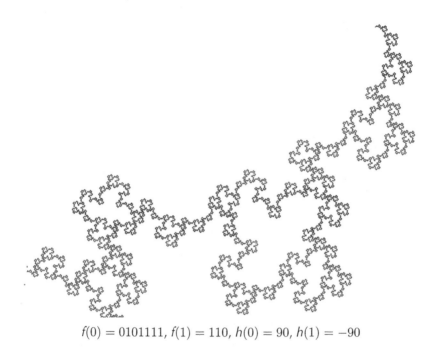

$f(0) = 0101111$, $f(1) = 110$, $h(0) = 90$, $h(1) = -90$

Here the figure has rotated many more turns. The starting point $(0, 0)$ appears at the top right. In this picture $P(f^7(0))$ is somewhere at the bottom. The point $P(f^8(0))$ is outside the picture. On this scale, the right angles from which the figure has been composed are hardly recognized any more.

Many variations on this idea are possible, even with non-right angles. For $f(0) = 010$, $f(1) = 10000010$, $h(0) = 72$, $h(1) = -72$ it is easy to check that the conditions of Theorem 11.1 hold; an initial part of the fractal turtle figure of $f^\infty(0)$ is the following:

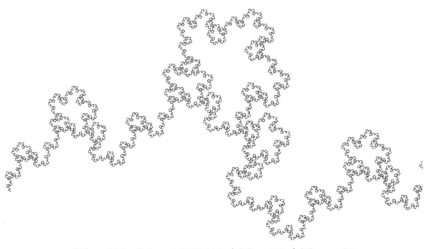

$$f(0) = 010, \ f(1) = 10000010, \ h(0) = 72, \ h(1) = -72$$

Here it is clear that the fractal turtle figure starts on the left. That is not always so obvious. For $f(0) = 011111010$, $f(1) = 1010$, $h(0) = 72$, $h(1) = -144$ it is again easy to verify that the conditions of Theorem 11.1 hold. An initial part of the fractal turtle figure of $f^\infty(0)$ is the following.

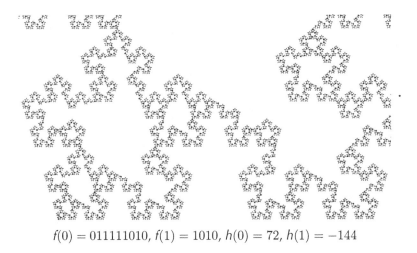

$$f(0) = 011111010, \ f(1) = 1010, \ h(0) = 72, \ h(1) = -144$$

The starting point is at the bottom, left in the bottom of the large pentagon.

In a variant we leave $f(1) = 1010$, $h(0) = 72$, $h(1) = -144$ unchanged, only the group of five ones in $f(0)$ is replaced by choosing $f(0) = 001111110$. This yields the following fractal turtle figure of $f^\infty(0)$:

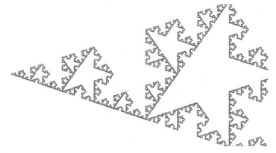

$$f(0) = 001111110, f(1) = 1010, h(0) = 72, h(1) = -144$$

Examples with $u = u'$

In the examples we've seen so far, we always applied Theorem 11.1 to words u and u' that were very similar, but not equal. We made one out of the another by inserting a word s^k somewhere for a number k for which $kh(s)$ was a multiple of 360°, so actually 0, hence still satisfying the condition $P(u) = P(u')$. A much simpler way to satisfy the condition $P(u) = P(u')$ is to simply choose $u = u'$. In this section, we give two examples of this, producing nice fractal patterns. In the first example, we choose $u = u' = 0010$, so $f(0) = 00010$ and $f(1) = 10010$. Since $u = u'$ contains three zeros and a 1, we need $3(h(0)) + h(1) = 0$ in order to satisfy the conditions of Theorem 11.1. The first example shows an initial part of the fractal turtle figure of $f^\infty(0)$ for $h(0) = 55$ and $h(1) = -165$.

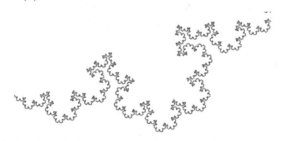

$$f(0) = 00010, f(1) = 10010, h(0) = 55, h(1) = -165$$

We get one of the most beautiful fractal turtle figures in a similar way, for which again the conditions of Theorem 11.1 are easily verified. Here we choose $u = u' = 00110$, so $f(0) = 000110$ and $f(1) = 100110$, and $h(0) = 70$ and $h(1) = -105$. The turtle starts on the right. From that on, pairs of trees appear, getting bigger and bigger to the left, and turning slightly counterclockwise.

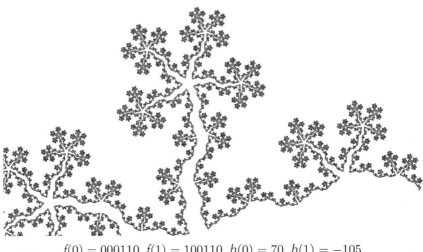

$$f(0) = 000110, f(1) = 100110, h(0) = 70, h(1) = -105$$

Now we switch to solutions of challenges. Contrary to what may be expected, this is about forthcoming challenges, more precisely, the challenges of the last two chapters. The reason for doing it here, earlier in the book than the challenge itself is the following. In the introductory chapter, it was promised to conclude every chapter with a challenge and also to present a solution to every challenge in this book. So it is inevitable that the solution of the last challenge will come before the challenge itself, otherwise the challenge is not the conclusion of that last chapter. So now we will give the solution of the challenge of the last Chapter 14. And in order not to make finding the solution to the Chapter 13 challenge too easy, we will include that solution here as well. The reader who has not yet seen these last two challenges should skip the following paragraphs for now.

The challenge of Chapter 13, is about creating a morphic sequence $a = f^\infty(0)$ whose proportion of zeros is exactly one percent. Stated in a formula that reads $[a]_0 = \frac{1}{100}$. We now show that this is indeed possible. Where that challenge is presented, Theorem 13.3 was just presented, and it is a good idea to use it for this challenge. That theorem states that if exactly one x exists such that $0 \leq x \leq 1$ and $(A + C - B - D)x^2 + (2B + D - A)x - B = 0$, where $A = |f(0)|_0, B = |f(1)|_0, C = |f(0)|_1$ and $D = |f(1)|_1$, then $[a]_0 = x$. So we need to choose A, B, C, D such that $x = \frac{1}{100}$ is the only solution satisfying $0 \leq x \leq 1$ of that equation. The easiest way to do that is to come up with the equation $100x - 1 = 0$, then we get $A + C - B - D = 0$, $2B + D - A = 100$ and $B = 1$. From $B = 1$ it follows that $A + C - D = 1$ and $D - A = 98$. These requirements can be simplified to $B = 1$, $C = 99$ and $D - A = 98$. Keep in mind that the definition of a purely morphic sequence requires $f(0)$ to start in 0, so A should satisfy $A > 0$. The simplest choice is then $A = B = 1$ and $C = D = 99$. So a possible choice is $f(0) = f(1) = 01^{99}$. This is correct, but produces the periodic sequence $a = (01^{99})^\infty$. Another non-periodic example is obtained by the same argument by choosing $f(0) = 01^{99}$ and $f(1) = 1^{99}0$.

The final challenge, from Chapter 14, starts in $A = 2B$, where among the 12 divisors of $B < 100$, there are exactly two primes and four squares. So B is of the shape $B = p^k \times q^n$ for two distinct prime numbers p, q and $k, n > 0$. The divisors of this number are exactly the numbers $p^{k'} \times q^{n'}$ with $0 \le k' \le k$ and $0 \le n' \le n$, of which there are exactly $(k + 1) \times (n + 1)$. Since the total number of divisors is 12 we obtain $(k + 1) \times (n + 1) = 12$. Let's assume that $k \le n$, otherwise we swap p and q, then this is only possible by $k + 1 = 2$ and $n + 1 = 6$ or by $k + 1 = 3$ and $n + 1 = 4$. In the first case, the number is of the shape $p \times q^5$, but then only three divisors are squares, namely 1, q^2 and q^4. So $k + 1 = 3$ and $n + 1 = 4$, and the number is of the shape $p^2 \times q^3$. The only instance of this below 100 is $72 = 3^2 \times 2^3$, because $2^2 \times 3^3 = 108$, and if any of the prime numbers ≥ 5 the result will be greater than 100. The number 72 indeed has four squares as divisors: 1, 4, 9, and 36. The six remaining divisors that are not a prime or a square are 6, 8, 12, 18, 24, 72. So we conclude that the process starts in $B = 72$ and $A = 144$. Then the assignment continues to the remark just before the googolplex challenge. There it was stated to add B to A and move on to the first page of the chapter on more complicated sequences. There the command was given to increase A by 10, only if B is even. Then the next instruction is found just before the challenge on knight moves. There, if A is even, B will be reduced by 1, and the process continues until just before the googolplex problem. We had already been there before, and so these three parts of the assignment will repeat forever. Such infinite behavior will be no surprise in this book on infinity. More precisely, the following three commands are repeated forever:

$A := A + B;$
if B is even, then $A := A + 10;$
if A is even, then $B := B - 1.$

Observing these commands, we see that B never increases, while sometimes it decreases. If the latter happens very often then B becomes negative, and even smaller than -10. Then by the commands $A := A + B$, sometimes followed by $A := A + 10$, the value of A will decrease too. So at this point, the highest value of A will have been reached, and the challenge was to figure out this highest value. An easy way is to write these commands in a programming language like Python, run the program, and see what comes out. I also did that as an extra check, but it is also possible to find this value without computer support, as we will now see. It is difficult to find a pattern in changing the values of A and B because some commands depend on whether A or B are even. So let's see what happens if we repeat running the three commands starting from a situation where A and B are both even. Let's call these starting values a and b. By the first three commands, A is replaced by $a + b + 10$, still being even, and B is replaced by $b - 1$, being odd. Again executing the three commands yields $a + 2b + 9$ for A, and $b - 1$ for B, both being odd. Running the three commands for the third time yields $A = a + 3b + 8$ and $B = b - 2$, both being even. So if we run three times the three commands on even numbers A and B the effect is

$$A := A + 3B + 8; \quad B := B - 2,$$

much simpler, and now with no tests at all. With the starting values $A = 144$ and $B = 72$, this yields in the first n steps (each corresponding to 9 commands) for $n = 0, 1, 2$:

n	A	B
0	144	72
1	368	70
2	586	68

If we continue this until $n = 36$, we get $B = 0$, and A will be close to its maximum value. But we don't want to do all these 36 steps by hand, so we are looking for an alternative.

By *triangular numbers* we have already seen that when starting in $A = 0$, $B = 1$, and running

$$A := A + B; \quad B := B + 1$$

n times, at the end the value of A is equal to $\frac{1}{2}n^2 + \frac{1}{2}n$. The factors are slightly different, but the analogy is big enough to guess that in our question the value of A after n steps will be of the form is $pn^2 + qn + r$, for numbers p, q, r still to be determined. Using the values from our table yields

$$p \times 0 + q \times 0 + r = 144,$$

$$p \times 1 + q \times 1 + r = 368,$$

$$p \times 4 + q \times 2 + r = 586.$$

From these properties we derive $r = 144$, $p + q = 224$ and $4p + 2q = 442$. From the last two we obtain $2p = 442 - 2 \times 224 = -6$, so $p = -3$. Then using $p + q = 224$ gives $q = 227$. So if it's of this form, then the value of A after n steps is $-3n^2 + 227n + 144$. And yes: one may prove by induction that this is indeed the case.

So we don't have to elaborate all first 36 steps, corresponding to $3 \times 36 = 108$ times the three commands ourselves, but we directly conclude that after these 108 times the three commands the value of A is equal to $-3 \times 36^2 + 227 \times 36 + 144 = 4428$, and the value of B is equal to 0. To determine the precise maximum value of A, we should consider the separate commands rather than only groups of three times three. However, it is sufficient to look at the triples of commands. This is done in the following tables, where k is the number of triples of commands:

k	A	B
108	4428	0
109	4438	-1
110	4437	-1
111	4436	-2
112	4444	-3
113	4441	-3
114	4438	-4
115	4444	-5
116	4439	-5
118	4434	-6

k	A	B
0	144	72
3	368	70
6	586	68
$\vdots$	$\vdots$	$\vdots$

Due negativity of B, after these steps A will not reach the value 4444 or higher again. Thus, the highest value for A that occurs in this never-ending process is 4444, which is the answer for this final challenge.

This problem has been designed deliberately in such a way that it produces this nice number 4444, consisting of four fours. When I searched the internet to look for a special meaning, I saw that it is called the *angel number*. In an explanation, I even found the following text by which the discussion of this last challenge is concluded: *The 4444 angel number signals that you should not abandon hope no matter how lost you may feel. Times are hard, and many people are struggling, but seeing 4444 is your signal to double up and keep pushing because better times are ahead.*

Now we are happy to return to the topic of this chapter: fractal turtle figures.

Other examples

An important requirement in Theorem 11.1 to arrive at fractal step figures is $P(u) = P(u')$. Until now this has been met by choosing u and u' equal, or making one out of the other by adding a word s^k for a number k such that $kh(s)$ is a multiple of $360°$. However, there are many more ways to fulfill this requirement. We will not investigate this exhaustively, but focus on a single suitable pattern. In case $h(1) = -h(0)$, then one easily checks that $P(0110) = P(1001)$ and $h(0110) = 0 = h(1001)$ hold. Below we see this for $h(0) = 50$, $h(1) = -50$:

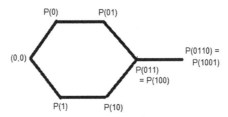

Notice that the path $(0,0)$, $P(0)$, $P(01)$, $P(011)$, $P(0110)$ is similar to the path $(0,0)$, $P(1)$, $P(10)$, $P(100)$, $P(1001)$, only mirrored over the horizontal x axis. The y- value of $P(0110) = P(1001)$ equals 0. This holds for all angles, as long as $h(1) = -h(0)$ holds. So Theorem 11.1 may be applied for $u = 0110$ and $u' = 1001$, or vice versa, and $h(1) = -h(0)$. The rotation angle is then 0 because the y value of $P(u) = P(u')$ equals 0. The scaling factor depends on the angle $h(0)$. We now present a number of examples.

For $f(0) = 00110$, $f(1) = 11001$, $h(0) = 90$ and $h(1) = -90$ we obtain the following fractal turtle figure, having rotation angle 0 and scaling factor 3. The figure starts at the bottom left. You may clearly recognize figures that appear again magnified by a factor 3, without rotation.

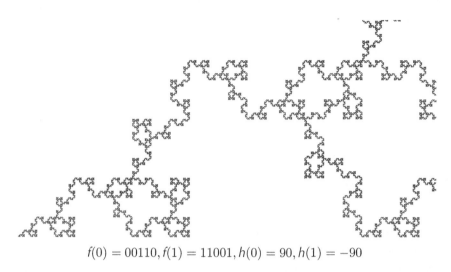

$$f(0) = 00110, f(1) = 11001, h(0) = 90, h(1) = -90$$

Using the same $f(0) = 00110$, $f(1) = 11001$, but now choosing $h(0) = 108$ and $h(1) = -108$ gives the following fractal turtle figure. Here the rotation angle is again 0, but the scaling factor is a number close to $2\frac{1}{2}$:

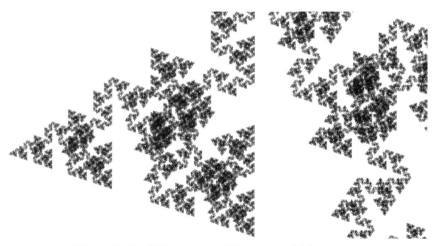

$$f(0) = 00110,\ f(1) = 11001,\ h(0) = 108,\ h(1) = -108$$

Now we swap u and u', so $f(0) = 01001$, $f(1) = 10110$. We again choose $h(0) = 90$ and $h(1) = -90$, and rotate the starting direction $45°$. On a rough scale, the initial part looks like this:

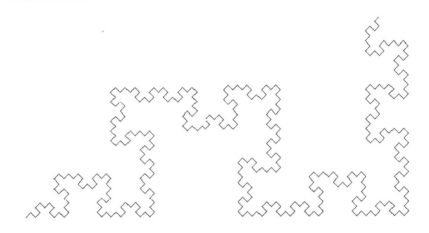

This fractal turtle figure again has rotation angle 0 and scale factor 3; it starts at the bottom left. In a finer scale, we show a much larger initial part. Here the turning of the starting direction of 45° is hardly visible any more.

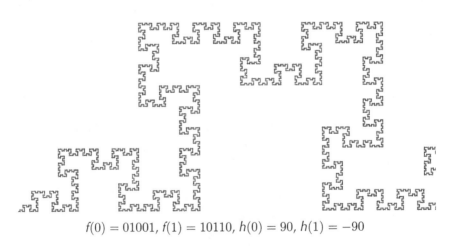

$$f(0) = 01001, f(1) = 10110, h(0) = 90, h(1) = -90$$

Finally, we give the fractal turtle figure for the same $f(0) = 01001, f(1) = 10110$, but now choosing $h(0) = 105$ and $h(1) = -105$.

$f(0) = 01001$, $f(1) = 10110$, $h(0) = 105$, $h(1) = -105$

Challenge: googol

We started this book by asking what the largest number is and soon figured out that for every number there exists a bigger one. But now let's make a really big number: the digit 1 followed by 100 zeros. This number may be written as 10^{100} and is called *googol*. Indeed it is very large. The name of the Google search engine was derived from this. Some people say it was a spelling error, others say it was spelt differently on purpose. Anyway: googol is very big. To give you an impression, the number of atoms in a human body is about 10^{28}, the number of atoms on the earth is about 10^{50}, and the number of atoms in the observable universe, that is, covering all galaxies that can be observed, is estimated to be at most 10^{82}. And that's one billionth of one billionth of googol, so still a lot smaller than googol.

Now we will formulate a challenge featuring this huge number googol. Consider words over the two letters a and b, and consider one simple rule: if the word contains the pattern ab, it may be replaced by baa. In doing so, the length of the word is increased by 1 by every step. So starting by the word *babaab* three consecutive steps are

$$babaab \rightarrow bbaaaab \rightarrow bbaabaa \rightarrow bbaabaaa.$$

Challenge:

Is it possible to start in a word consisting of less than 1000 letters *a* and *b*, and then do googol steps according to this simple rule, and end up in a word of more than googol letters?

12 | **Variations on Koch**

Koch curve

One of the first fractal figures in literature is described in a 1904 paper by the Swedish mathematician Helge von Koch (1870–1924). It is therefore known as the *Koch curve*. The idea is as follows: start by a horizontal segment of length 1 and replace it with the following pattern consisting of four segments of length $\frac{1}{3}$:

The distance between the ends remains 1. Then the same is repeated on these four line segments: each of them is replaced by the same pattern, shrunk by a factor of 3, leaving the ends of the entire figure in place:

Doing this again by replacing each of these 16 line segments with four new ones yields the following figure of 64 line segments:

DOI: 10.1201/9781003466000-12

Doing it once more yields the following:

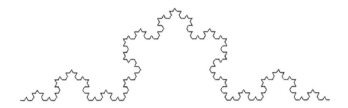

Finally, the Koch curve is defined as the limit of this process. The closer we get to that limit, the shorter the line segments. On a slightly larger scale, it looks like this:

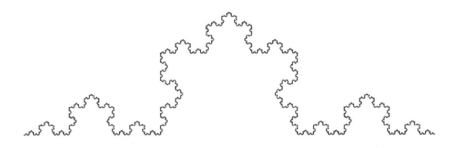

In this process, in every step, each line segment is replaced by four line segments, each being three times smaller than the line segment itself. We started with only one line segment of length 1. After one step, there were four line segments of length $\frac{1}{3}$, having total length $\frac{4}{3}$. After two steps, the total length is again multiplied by $\frac{4}{3}$, yielding $(\frac{4}{3})^2$. In repeating this, for every n it holds that after n steps there are 4^n line segments that each have length $(\frac{1}{3})^n$, having a total length of $(\frac{4}{3})^n$. But in the limit, in the final curve, this is done infinitely many times. As n goes to infinity, the total length $(\frac{4}{3})^n$ also goes to infinity. So the Koch curve has infinite length, no longer has a direction anywhere, but still operates within a bounded area, and connects two points of distance of 1 from each other.

Instead of starting in a segment, it is also possible to start in a triangle:

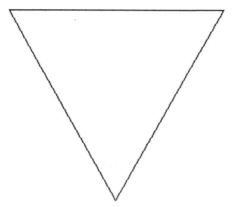

Then performing the above process actually yields three copies of Koch's curve, together forming a closed figure:

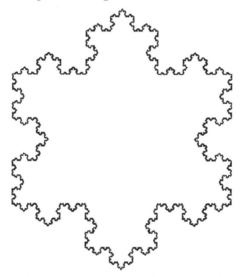

This figure is known as the *Koch snowflake*.

While the circumference of the Koch snowflake is infinite, its area is not, because everything takes place within a bounded area. We calculate that area as follows.

Let's write A for the area of the original triangle. The side of the triangle is 1, the height is $\frac{1}{2}\sqrt{3}$, so we have $A = \frac{1}{4}\sqrt{3}$. After the first step, three triangles are added having sides $\frac{1}{3}$, each having area $\frac{A}{9}$, so together they have an area $\frac{A}{3}$. In each subsequent step, four times as many triangles are added, each having area nine times smaller than the previous series of triangles. That gives the total surface area

$$A(1 + \frac{1}{3}(1 + \frac{4}{9} + (\frac{4}{9})^2 + (\frac{4}{9})^3 + (\frac{4}{9})^4 + \cdots)).$$

Write P for $1 + \frac{4}{9} + (\frac{4}{9})^2 + (\frac{4}{9})^3 + \cdots$. Then this formula yields $P = 1 + \frac{4}{9}P$, from which it follows that $P = \frac{9}{5}$. If we plug that into the formula above, we see that the total area is $A(1 + \frac{3}{5})$. So the area of the Koch snowflake is $\frac{8}{5}$ times the area of the original triangle, so is $\frac{2}{5}\sqrt{3}$.

Koch curve as a turtle figure

Is it possible to present the Koch curve as a turtle figure? Let's start by a crucial difference. In the construction of the Koch curve, the line segments become increasingly smaller, while in a turtle figure the line segments always keep a fixed length. However, all finite approximations of the Koch curve turn out to be turtle figures of words, and every subsequent approximation is again a turtle figure of an extended word. Continuing all these extensions then gives an infinite sequence, which will now be presented as a purely morphic sequence.

The first approach

is drawn from left to right by bending 60° to the left after the first step, 120° to the right after the second step, and once more 60° to the left after the third step. So that is the turtle figure of 010 for the angles $h(0) = 60$ and $h(1) = -120$. This is what happens after the first step, it should be preceded by this initial step. To create a morphic sequence $f^\infty(0)$ the word $f(0)$ should start in 0. So by choosing $f(0) = 0010$, this first approximation is obtained as the turtle figure of $f(0)$. The initial direction may be chosen in such a way that the first segment is horizontal. For the turtle figure, the effect of replacing every 0 by 0010 means that every line segment belonging to 0 is replaced by this first approximation and becomes three times as long. Similarly, by replacing every 1 by 1010, every segment associated with a 1 is replaced by this same pattern and is made three times as long. This is how we choose the f: $f(0) = 0010$ and $f(1) = 1010$. With the already chosen $h(0) = 60$ and $h(1) = -120$ the second approximation then becomes the turtle figure of the word $f(f(0)) = 0010001010100010$, being

And this continues: the turtle figure of $f^3(0)$ is exactly the third approximation

and so on in every next step.

If we now choose a large value of n, and we scale the step figure of $f^n(0)$ in such a way that the unit step is no more than one pixel, the result is exactly Koch's curve. Formally, we should go on approximating indefinitely, but assuming we can't see smaller than a pixel anyway, then there's no difference between what we think of as the Koch curve and the nth approximation. So one way to represent the Koch curve in any precision is to choose n large enough and make the turtle figure of $f^n(0)$. The ultimate turtle figure of $f^\infty(0)$ goes on indefinitely, so if we want to show it, it is unavoidable to stop it somewhere, and a natural point to break off is just $f^n(0)$.

So in this way, the turtle curve of the sequence $f^\infty(0)$ for the morphism f defined by $f(0) = 0010$ and $f(1) = 1010$, and the given angles, coincides with the Koch curve. Hence, we call this sequence the *Koch sequence*, and write it as

$$\mathbf{k} = f^\infty(0) = 0010001010100010\cdots.$$

We call the turtle figure of $\mathbf{k}$ at the given angles $h(0) = 60$ and $h(1) = -120$ the *Koch turtle figure*. On a fine scale, we can see it as an infinite continuation of the Koch curve.

Due to the fractal behavior of the Koch curve, it is no surprise that this Koch turtle figure is fractal. To prove this, we don't have to do any extra work at all: this follows directly from Theorem 11.1: by choosing $u = u' = 010$ we obtain $h(u) = 0$ and $P(u) = (2,0)$, and according to the theorem, the turtle figure is fractal, having rotation angle 0 and scaling factor 3.

Other scaling factors

We saw that for the angles $h(0) = 60$ and $h(1) = -120$ the Koch sequence yields a fractal turtle figure according to Theorem 11.1, having rotation angle

0 and scaling factor 3. But the conditions of the same theorem also apply to other angles as long as $2h(0) + h(1) = 0$. Then the rotation angle is still 0, but the scale factor is no longer 3. More precisely, the following figure shows why $P(u) = P(010) = (1 + 2\cos(h(0)), 0)$, so according to Theorem 11.1 the turtle figure is fractal, having rotation angle 0, and scaling factor $2 + 2\cos(h(0))$. Here we again use the *sine* and *cosine* of angles as we have seen before.

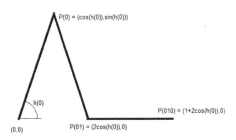

Indeed, in the Koch turtle figure, we have $h(0) = 60$, giving $\cos(h(0)) = \frac{1}{2}$, so the scaling factor is $2 + 2\cos(h(0)) = 3$.

If we choose the angle $h(0)$ to be slightly below 90°, then $\cos(h(0))$ is just over 0, so the scaling factor is just over 2. The resulting fractal figure for $h(1) = -2h(0)$ is then called the *Caesàro fractal*, named after Ernesto Cesàro (1859–1906). For instance, $h(0) = 87$ and $h(1) = -174$ gives the following fractal turtle figure, being an example of the Caesàro fractal:

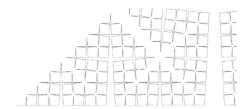

Here we chose a quite coarse scale. The same figure on a finer scale looks as follows. Again it starts on the left and continues indefinitely to the right and top right. Again only a finite part is shown, but in this case, it involves more than half a million elements from the Koch sequence:

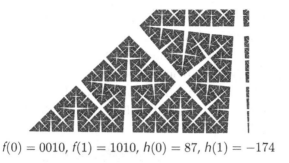

$f(0) = 0010$, $f(1) = 1010$, $h(0) = 87$, $h(1) = -174$

A nice observation is that in this way we obtain fractal turtle figures having rotation angle 0 and scaling factor $2 + 2\cos(h(0))$. By choosing suitable $h(0)$, this scaling factor $2 + 2\cos(h(0))$ may be any real number between 0 and 4.

If we choose $h(0)$ very small, for example a few degrees, then this scaling factor is almost 4. This does not produce exciting turtle figures: they are close to being straight lines only having some very slight bends in them. If we make $h(0)$ bigger, it starts to resemble Koch's curve more and more, while at $h(0) = 60$ we end up exactly there. If $h(0)$ continues to grow, it remains very similar to the Koch curve at first, and if $h(0)$ approaches 90°, we get the Caesàro fractal.

When choosing $h(0)$ to be exactly 90°, so $h(1) = -180 = 180$, in a fine scale it becomes a 45° angle that is completely filled black. And on a somewhat coarser scale it just shows a rectangular grid. In this case the scaling factor is exactly 2, since $\cos(90) = 0$.

When choosing $h(0)$ more than 90°, the scale factor becomes <2, because $\cos(h(0))$ is then negative. Let's choose $h(0) = 105$, so $h(1) = -210 = 150$. After turning the starting direction, the corresponding turtle figure of the initial part $f(0) = 0010$ looks like this:

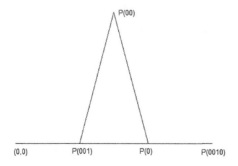

The first step is horizontally from $(0,0)$ to $P(0)$. Then we rotate 105° and take another step, to $P(00)$. Then we rotate $h(1) = -210 = 150°$ and go one step diagonally downwards, to $P(001)$. Finally, we rotate 105° again, bringing the direction back to horizontal, and take the last step to $P(0010)$. Note that the part between $P(0)$ and $P(001)$ is traversed twice. If we write d for the distance between $P(0)$ and $P(001)$, then the scaling factor is $2 - d$.

A greater initial part of the fractal turtle figure becomes quite a mess in this case. On a fine scale, big parts will be completely black. On a coarse scale, the initial part looks like this:

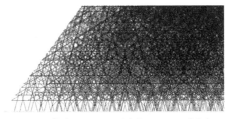

$$f(0) = 0010, \; f(1) = 1010, \; h(0) = 105, \; h(1) = 150$$

At the bottom left, the start with the just shown turtle figure of $f(0) = 0010$ can clearly be recognized as well as many enlarged copies.

When choosing $h(0) = 120$ and $h(1) = -240 = 120$ for both symbols 0 and 1 the turtle rotates over an angle of 120°, followed by a step. This yields an equilateral triangle for every sequence, and therefore also for **k**. Because of $\cos(120) = -\frac{1}{2}$ then the scaling factor is exactly 1. So theorem 11.1 does not apply because for being fractal it is required that the scaling factor c is unequal to 1.

It is interesting to note that for angles $h(0)$ greater than 120°, the scaling factor $c = 2 + 2\cos(h(0))$ becomes <1, because then $\cos(h(0)) < -\frac{1}{2}$. Then Theorem 11.1 does apply. It is surprising that this exists: a fractal turtle figure with a scaling factor smaller than 1, while it still purely consists of line segments of fixed length 1. Unfortunately, it gives no nice pictures as all these points close together only give big dirty black spots.

The period-doubling sequence

In this section, we present a sequence that is very similar to the Koch sequence, but has a simpler definition. It is defined as

$$\mathbf{p} = f^{\infty}(0) = 0100010101000100\cdots$$

for the morphism f defined by $f(0) = 01$ and $f(1) = 00$. Since $f(s)$ consists of two symbols for both s symbols, if a is a periodic sequence u^{∞} with period

n, then $f(a) = (f(u))^\infty$ also periodic, but with period $2n$. So the morphism f is *period-doubling* when applied to a periodic sequence. That is why $\mathbf{p}$ is called the *period-doubling sequence*. For the same reason, for any morphism f for which every $f(s)$ consists of two symbols, the associated morphic sequence $f^\infty(0)$ could be called period-doubling. So this holds for the Thue-Morse sequence too. However, in the literature, only $\mathbf{p}$ is called *the* period-doubling sequence, so let's do that too.

The connection between the Koch sequence $\mathbf{k}$ and this new sequence $\mathbf{p}$ is remarkable: the sequence $\mathbf{p}$ is obtained by simply removing the first element 0 from $\mathbf{k}$. We already denoted this removal of the first element of a sequence by tail. So this property is stated in the following theorem:

Theorem 12.1

$$\mathbf{p} = \text{tail}(\mathbf{k}).$$

Of course we should not just believe this theorem, but need a proof for it. To give this proof, we use the principle of *strong induction* as we discussed in the chapter on natural numbers. Let's recall this principle:

> If we can prove the property $P(n)$ for every n under the assumption that $P(i)$ holds for all $i < n$, then $P(n)$ holds for every natural number n.

The assumption that $P(i)$ holds for all $i < n$ is called the *induction hypothesis*.

Now we give the proof of Theorem 12.1.

Proof: We prove the theorem by proving that for every n holds: $\mathbf{p}_n = \mathbf{k}_{n+1}$. As announced we do this by strong induction: we define $P(n)$ to be the property $\mathbf{p}_n = \mathbf{k}_{n+1}$.

Before giving the induction proof we derive some basic properties of $\mathbf{p}$ and $\mathbf{k}$. Since $\mathbf{p} = f(\mathbf{p})$ for f defined by $f(0) = 01$ and $f(1) = 00$ we obtain $\mathbf{p}_{2i} = 0$ for every i, since for both values of s the first symbol of $f(s)$ is 0. Since the second symbol of $f(s)$ is $\neg s$ for both values of s, we obtain $\mathbf{p}_{2i+1} = \neg\mathbf{p}_i$ for every i. Here $\neg$ is defined by $\neg 0 = 1$ and $\neg 1 = 0$.

For $\mathbf{k}$ we do something similar. Since $\mathbf{k} = f(\mathbf{k})$ for f defined by $f(0) = 0010$ and $f(1) = 1010$, we obtain $\mathbf{k}_{4i+1} = \mathbf{k}_{4i+3} = 0$ for every i, because the second and fourth elements of $f(s)$ are 0 for both values of s. Similarly we obtain $\mathbf{k}_{4i+2} = 1$ for every i, because the third element of $f(s)$ is 1 for both values of s. Finally, $\mathbf{k}_{4i} = \mathbf{k}_i$ for every i, because the first element of $f(s)$ is always s for both values of s.

Now we come back to what we need to prove: $\mathbf{p}_n = \mathbf{k}_{n+1}$. We do some case analysis. If n is even then according to the above observations, we have $\mathbf{p}_n = 0$. But then $n + 1$ is odd, so according to the above observations we have $\mathbf{k}_{n+1} = 0$, because every odd number is of the form $4i + 1$ or $4i + 3$. So for n being even we're done. If n is odd, we have two cases: $n = 4i + 1$ or $n = 4i + 3$. If $n = 4i + 1$, then according to the above properties we have

$$\mathbf{p}_n = \neg\mathbf{p}_{2i} = \neg 0 = 1 = \mathbf{k}_{4i+2} = \mathbf{k}_{n+1},$$

exactly what we had to prove. In the last case, we have $n = 4i + 3$. According to the above properties, we then have

$$\mathbf{p}_n = \mathbf{p}_{2(2i+1)+1} = \neg\mathbf{p}_{2i+1} = \neg\neg\mathbf{p}_i = \mathbf{p}_i \text{ and } \mathbf{k}_{n+1} = \mathbf{k}_{4(i+1)} = \mathbf{k}_{i+1}.$$

Because $i < n$ we may apply the induction hypothesis on i: $\mathbf{p}_i = \mathbf{k}_{i+1}$. So also in this last case we have $\mathbf{p}_n = \mathbf{k}_{n+1}$, which completes the proof. $\square$

Fractal turtle figures of variants of p

By using the just proved property that the period-doubling sequence $\mathbf{p}$ is obtained from $\mathbf{k}$ by removing the first element, we see that turtle figures of $\mathbf{p}$ are obtained from turtle figures of $\mathbf{k}$ by removing the first line segment. In case the turtle figure of $\mathbf{k}$ is fractal, for example if $h(0) + 2h(1) = 0$, then so is the turtle figure of $\mathbf{p}$. This is not completely obvious and depends on the definition of *fractal*. Until now the point $(0, 0)$ fulfilled two roles: it was both the starting point of the turtle figure, and it was the point over which rotation took place for the fractal behavior. In all our examples so far they coincided, but now that is not the case any more. Let us fix $(0, 0)$ as the point of rotation. Then $(0, 0)$ is also the starting point of the turtle figure of $\mathbf{k}$. However, it is not the starting point of the turtle figure of $\mathbf{p}$, because that starts after the first line segment of the turtle figure of $\mathbf{k}$. From now on, we do not require in the definition of fractal turtle figure that these two roles are represented by the same point. Hence a fractal turtle figure of $\mathbf{k}$ always yields a fractal turtle figure of $\mathbf{p}$.

In a fine scale, no difference is visible at all between the turtle figures of $\mathbf{k}$ and $\mathbf{p}$ because they only differ in that single line segment at the beginning. So it makes no sense to show a picture of a turtle figure of $\mathbf{p}$ for angles for which we already showed the turtle figure of $\mathbf{k}$, because they only differ in the very first line segment.

But what does give surprises is applying the trick of adding a word s^n in the definition of $f(0)$ or $f(1)$ for which $nh(s)$ is a multiple of $360°$. If the origin was fractal, then the result of this adjustment remains fractal, in a similar way to Theorem 11.1. We do not elaborate the detailed proof here; it is given in my paper that appeared in the journal *Fractals* in 2016. Instead here we give several examples.

In the first example, we choose $h(0) = 45$ and $h(1) = -90$. This yields a fractal turtle figure because of $h(0) + 2h(1) = 0$. Since $4h(1) = -360 = 0$ we are allowed to add four ones in the definition of f, and so we do: we choose $f(0) = 011111$ and $f(1) = 00$. We rotate the starting direction so that the fractal turtle figure grows from left to right. On a rough scale, the initial part looks like this:

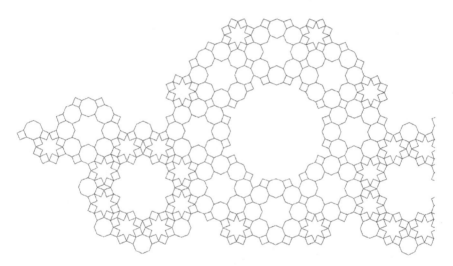

Because of the angles of $45°$ and $-90°$, several recognizable squares and regular octagons appear. Also the fractal behavior is nicely observed: the square and the octagon that started at the far left continuously create enlarged copies, filled by copies of themselves. This fractal behavior is even better visible in a much larger initial part of the same turtle figure on a much finer scale:

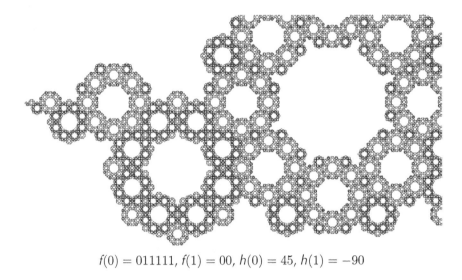

$$f(0) = 011111, f(1) = 00, h(0) = 45, h(1) = -90$$

As this example is based on the period-doubling sequence coinciding with the Koch sequence up to its first element, the rotation angle is 0, just like in the Koch turtle figure. Also the scaling factor is the same as for the Koch sequence, and is $2 + 2\cos(45) = 2 + \sqrt{2}$, being approximately 3.41.

By choosing $h(0) = 36$ and $h(1) = -72$ the same can be done, then we have $5h(1) = -360 = 0$, so we have to add five ones to the definition of **p**. For $f(0) = 0111111$ and $f(1) = 00$ we give the resulting fractal turtle figure:

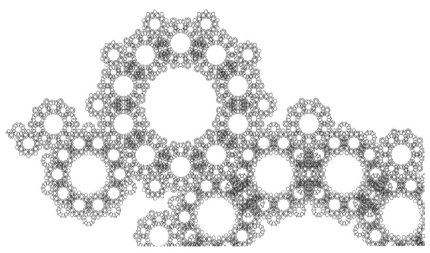

$$f(0) = 0111111, f(1) = 00, h(0) = 36, h(1) = -72$$

The same argument gives a fractal turtle figure for the same morphism f at the angles $h(0) = 72$ and $h(1) = -144$:

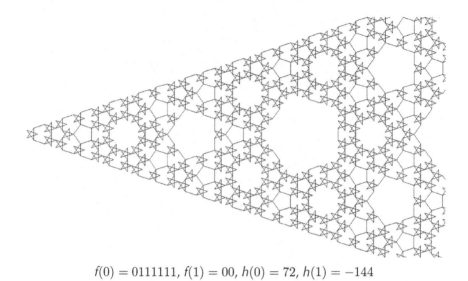

$$f(0) = 0111111, \ f(1) = 00, \ h(0) = 72, \ h(1) = -144$$

Due to this angle $h(1) = -144$, every occurrence of 11111 yields a five-pointed star, so we see many of these stars in this pattern. A much larger initial part and a much finer scale give the fractal turtle figure that we gave as an example in the introductory chapter, rotated over 90°. There the individual stars are very small, but the ever-expanding pentagons appear clearly.

Right after that in the introductory chapter, we gave a large filled pentagon. That was also the initial part of a fractal turtle figure having the same angles $h(0) = 72$ and $h(1) = -144$, but now for $f(0) = 01$ and $f(1) = 1111100$. So the pattern 11111 is not added at the end of $f(0)$ but at the beginning of $f(1)$. On a coarser scale, a much smaller initial part consisting of the first 35,000 elements of the sequence gives the following turtle figure.

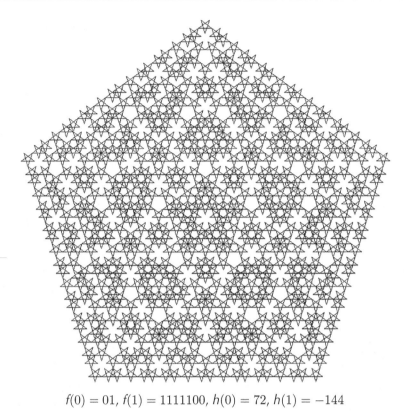

$$f(0) = 01, f(1) = 1111100, h(0) = 72, h(1) = -144$$

We just gave a series of examples in which five ones were added in $f(0)$ or $f(1)$, and the angle $h(1)$ was chosen such that $5h(1)$ is a multiple of $360°$. The same trick works for other numbers. Now we give an example where we add six ones for an angle $h(1)$ with $6h(1)$ a multiple of $360°$. To be precise: $f^\infty(0)$ for $f(0) = 01111111 = 01^7$ and $f(1) = 00$ yields the following fractal turtle curve for the angles $h(0) = 30$ and $h(1) = -60$. It starts on the left, slightly below the center, and continues indefinitely to the right.

$$f(0) = 01111111, \; f(1) = 00, \; h(0) = 30, \; h(1) = -60$$

Please look at this figure in detail and observe all kinds of fractal behavior. It is surprising to see how all this originates from the very simple definition of f, and the far from exotic angles of 30° and 60°, not having any apparent fractal behavior at a first glance.

The following sentence is only intended for those who are working on the challenge of the last chapter. The number B is added to A, and then the process continues by an instruction on the first page of the chapter on more complicated sequences.

Challenge: googolplex

In the previous challenge, we encountered the number *googol*. That is the digit 1 followed by 100 zeros, so it is 10^{100}. We observed that this number is much more than an estimates of the total number of atoms in the universe. While googol is already much larger than any number we will ever encounter as a physical number of anything, it is possible to go much further. The number *googolplex* is defined to be $10^{\text{googol}} = 10^{10^{100}}$. So in the usual notation, it would be the digit 1 followed by googol zeros: definitely impossible to write down in this notation. Try to imagine how terribly big that number is. This number googolplex will play a role in the new challenge that will be presented now.

To get a feeling on how to deal with such big numbers, let us first return to the previous challenge. The question was: by only replacing *ab* by *baa*, is it possible to perform googol steps when starting in a word having <1000 symbols? When starting in $a^n b$ for any number n, in every step the *b* shifts one position to the left and a fresh symbol *a* is added to the right, so in n steps the word $a^n b$ is converted to ba^{2n}. By repeating this observation, and starting at ab^n for some number n we obtain the following conversion:

$$ab^n \rightarrow baab^{n-1} \rightarrow bba^{2 \times 2}b^{n-2} \rightarrow bba^{2 \times 2 \times 2}b^{n-3} \rightarrow \cdots \rightarrow b^n a^{2^n}.$$

Because the length of the word is increased by $2^n - 1$ in this conversion, the number of steps is $2^n - 1$, and this number can get really big now. The challenge was about starting in a word of <1000 symbols. Let's choose ab^{400} of length 401, definitely <1000. By the above observation, this yields $2^{400} - 1$ steps. How about the size of 2^{400}? That equals $(2 \times 2 \times 2 \times 2)^{100} = 16^{100}$, being much larger than 10^{100}, which is the number googol. So starting in the word ab^{400} of length 401 it is possible to do much more than googol steps. When stopping after googol steps, it shows that it is possible to do exactly googol steps. This completes the previous challenge.

For the new challenge, we are going to play the following game with six boxes in a row, which may be filled with marbles. We assume that the number of spare marbles is unbounded, not worrying about the issue that having googol marbles or more will be physically impossible. We consider steps according to the following rules:

- If one of the first five boxes is not empty, we may move one of its marbles to its right-hand neighbor box, moreover adding one extra marble from the spare supply to that box. So if those two neighboring boxes contain i and j marbles, after this step this will be $i - 1$ and $j + 2$.
- If one of the first four boxes is not empty, we may remove one marble from it, that is, move it to the spare supply, while the two boxes to the right of it are swapped. So if the three boxes involved contain i, j and k marbles, then after this step this will be $i - 1$, k and j.

For example, if the six boxes contain $3, 2, 1, 7, 2, 5$ marbles, then after a step of the first type, this may be transformed to $2, 4, 1, 7, 2, 5$, and then after a step of the second type, this may be transformed to $2, 3, 7, 1, 2, 5$.

Challenge:

Is it possible to do infinitely many steps according to the above rules, when starting in a configuration with finitely many marbles?

Now consider the configuration in which the first two boxes both contain one marble, and the other four boxes are empty. Is it possible to obtain more than googolplex marbles in one of the boxes by starting in this configuration and only applying the above rules?

Simple morphic sequences

This chapter investigates morphic sequences of the very simplest kind: purely morphic sequences $f^\infty(0)$ over only two symbols 0 and 1 in which $f(0)$ and $f(1)$ both consist of at most two symbols.

We have already seen two of these: the Thue-Morse sequence **t** defined as

$$\mathbf{t} = f^\infty(0) = 0110100110010110\cdots$$

for f defined by $f(0) = 01$ and $f(1) = 10$, and the period-doubling sequence **p** defined as

$$\mathbf{p} = f^\infty(0) = 0100010101000100\cdots$$

for f defined by $f(0) = 01$ and $f(1) = 00$.

It turns out that these two sequences are closely related. Some of this relationship will be discovered by considering some more turtle figures of the Thue-Morse sequence **t**, as we will see now.

Koch-like turtle figures of Thue-Morse

Most of the examples of turtle figures we gave until now were created by first deriving conditions on morphisms and angles in order to obtain some desired properties, like being finite or fractal, and then choose morphisms and angles satisfying these conditions. Now we do it the other way around: we choose a very simple morphic sequence, namely the Thue-Morse sequence **t**, and arbitrarily some simple angles $h(0)$ and $h(1)$, and see how the resulting turtle figures look like. If we choose the angles such that $h(0) + h(1)$ is a multiple of 360° divided by a power of 2, then we know by Theorem 9.3 that the resulting turtle figure is finite. We have seen many examples of this. But what

DOI: 10.1201/9781003466000-13

if we try 60°: a very basic angle not having this property? Let's choose a most simple instance: $h(0) = 60$ and $h(1) = 0$. An initial part of the turtle figure then looks like this:

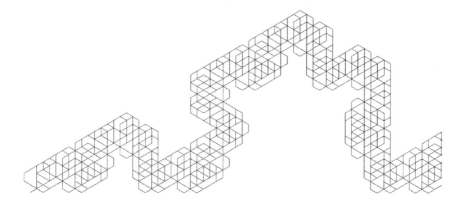

At first glance, you may see a kind of three-dimensional block pattern. But now let's take a larger initial part in a much finer scale:

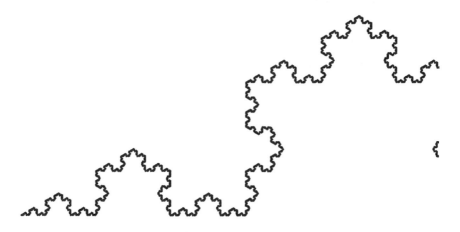

Hey, this we do recognize! It looks like the Koch curve, only fattened up a bit. Is that a coincidence, or is there some reason why this happens?

Let's consider some more variants. We keep the same Thue-Morse sequence **t**, but modify the angles $h(0)$ and $h(1)$, maintaining $h(0) + h(1)$ to be some multiple of 60°, but not a multiple of 180°. It turns out that again the same phenomenon shows up: on a coarse scale, it may be different, but on a fine scale, it looks like a fattened Koch curve every time. We give two examples. First we choose $h(0) = 180$ and $h(1) = 60$. The turtle figure then looks like this:

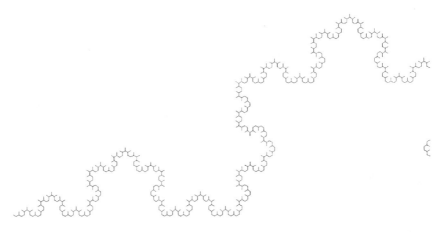

Again the Koch curve is clearly recognizable, but when zooming in it is different. Because of $h(0) = 180$, at every 0 the last drawn line segment is redrawn in the reverse direction, which is why you see those dead-end line segments everywhere.

In the following example we choose $h(0) = 180$ and $h(1) = 120$. The turtle figure then is as follows:

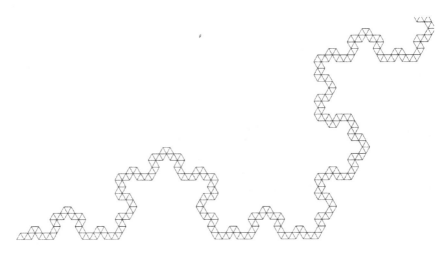

Here it consists of many triangles glued together, together forming a figure very similar to the Koch curve. For these last two examples, a representation of a larger initial part in a finer scale could be given too, but we do not do that here: in both cases we again get a much more precise representation of the Koch curve, only being fattened as before.

Are these turtle figures fractal?

No, they are not. If they were, then every pattern should repeatedly return enlarged. In this last example, we see a lot of small triangles, but no enlarged copies of them. So this turtle curve does not meet our definition of fractal turtle curve, and the same holds for the other examples.

The figures suggest that there is a strong connection with the Koch curve. Indeed there is. To be precise, by connecting the midpoints of the $16n$th segment of the last turtle figure for $n = 1, 2, 3, \ldots$, the Koch curve is obtained exactly:

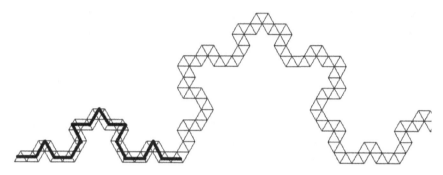

Here also the connection between the midpoints of the first and the 16th line segment is indicated to get the full Koch curve. Because $\mathbf{p}$ is obtained from the Koch sequence by removing the first element, the Koch curve without this first line segment is obtained as the turtle figure of $\mathbf{p}$ for $h(0) = 60$ and $h(1) = -120$. So this turtle figure is also obtained by taking the turtle figure of $\mathbf{t}$ for $h(0) = 180$ and $h(1) = 120$, and sequentially connecting the midpoints of the $16n$th line segments for $n = 1, 2, 3, \ldots$ The full proof is omitted here. Instead, we give a very strong relationship between $\mathbf{t}$ and $\mathbf{p}$ as sequences.

Relating t and p

We give a theorem showing the relation between the Thue-Morse sequence $\mathbf{t}$ and the period-doubling sequence $\mathbf{p}$. More precisely, we show how for every n the n-th element of $\mathbf{p}$ is obtained from two subsequent elements of $\mathbf{t}$.

Theorem 13.1 *For any natural number n, we have $\mathbf{p}_n = 0$ if $\mathbf{t}_n$ and $\mathbf{t}_{n+1}$ are distinct, and $\mathbf{p}_n = 1$ if $\mathbf{t}_n$ and $\mathbf{t}_{n+1}$ are equal.*

Proof: As in the proof of Theorem 12.1, we use strong induction. From the proof of Theorem 12.1 we already know that $\mathbf{p}_{2i} = 0$ and $\mathbf{p}_{2i+1} = \neg\mathbf{p}_i$ for every i.

In the chapter on the Thue-Morse sequence we already gave the following property of $\mathbf{t}$: for every i we have $\mathbf{t}_{2i} = \mathbf{t}_i$ and $\mathbf{t}_{2i+1} = \neg\mathbf{t}_i$.

We split the proof of the theorem into two cases: n even and n odd. If n is even, then $n = 2i$, then $\mathbf{p}_n = 0$ and

$$\mathbf{t}_n = \mathbf{t}_{2i} = \neg\neg\mathbf{t}_i = \neg\mathbf{t}_{2i+1} = \neg\mathbf{t}_{n+1}.$$

So $\mathbf{t}_n$ and $\mathbf{t}_{n+1}$ are distinct, exactly what we had to prove.

In the remaining case, n is odd. Then we may write $n = 2i + 1$, and we have $\mathbf{p}_n = \neg\mathbf{p}_i$, $\mathbf{t}_n = \mathbf{t}_{2i+1} = \neg\mathbf{t}_i$ and $\mathbf{t}_{n+1} = \mathbf{t}_{2i+2} = \mathbf{t}_{i+1}$. Since $i < n$ we may assume by the induction hypothesis that $\mathbf{p}_i = 0$ if $\mathbf{t}_i$ and $\mathbf{t}_{i+1}$ are distinct, and that $\mathbf{p}_i = 1$ if $\mathbf{t}_i$ and $\mathbf{t}_{i+1}$ are equal. If $\mathbf{t}_n = \neg\mathbf{t}_i$ and $\mathbf{t}_{n+1} = \mathbf{t}_{i+1}$ are distinct, then $\mathbf{t}_i$ and $\mathbf{t}_{i+1}$ are exactly equal, and according to the induction hypothesis, we have $\mathbf{p}_i = 1$, so $\mathbf{p}_n = \mathbf{p}_{2i+1} = \neg\mathbf{p}_i = 0$. Conversely, if $\mathbf{t}_n = \neg\mathbf{t}_i$ and $\mathbf{t}_{n+1} = \mathbf{t}_{i+1}$ are equal, then $\mathbf{t}_i$ and $\mathbf{t}_{i+1}$ are distinct, and according to the induction hypothesis, we have $\mathbf{p}_i = 0$, so $\mathbf{p}_n = \mathbf{p}_{2i+1} = \neg\mathbf{p}_i = 1$. This is exactly what we had to prove. $\square$

This strong relationship between $\mathbf{t}$ and $\mathbf{p}$ can be seen as an indication for the similarity of the turtle figures of $\mathbf{t}$ and $\mathbf{p}$.

Finite turtle figures

For $\mathbf{t}$ we have seen a lot of finite turtle figures. We did not see purely fractal turtle curves of $\mathbf{t}$, but only related to being fractal: by connecting the midpoints of certain line segments, we got exactly the Koch fractal turtle figure.

From $\mathbf{p}$ we have seen that it gives several fractal turtle figures, in particular the Koch figure. But does $\mathbf{p}$ also provide finite turtle figures?

Yes, it does! It turns out that in many cases Theorem 10.1 applies, giving conditions to conclude finiteness of a turtle figure. The argument we will give here is very similar to the argument we gave for the finiteness of the figures called *rosettes*. For any m to be chosen, the sequence $\mathbf{p}$ is composed from $f^m(0)$ and $f^m(1)$, for f defined by $f(0) = 01$ and $f(1) = 00$. We have to choose the angles such that $h(f^m(1)) = 0$ and $P(f^m(1)) = (0, 0)$, the most important condition for Theorem 10.1. The central observation is that this condition is satisfied if $h(f^{m-1}(0)) = 180$, up to a multiple of $360°$. This holds because

$$f^m(1) = f^{m-1}(f(1)) = f^{m-1}(00) = f^{m-1}(0)f^{m-1}(0).$$

If $h(w) = 180$ for a word w then $h(ww) = 360 = 0$ and $P(ww) = (0,0)$, the latter because after the first w the path is traversed again, but then rotated $180°$, bringing the turtle back to the starting point. We have seen this argument before, in particular for the rosettes. So for $w = f^{m-1}(0)$ the condition is met. In the examples we give, the other condition of Theorem 10.1 is also satisfied, that is, $h(f^m(0))$ is rational and unequal to 0.

In the first example we choose $h(0) = 36$ and $h(1) = 72$:

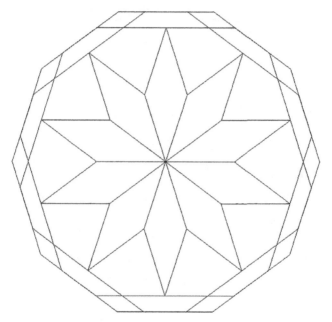

$$f(0) = 01, f(1) = 00, h(0) = 36, h(1) = 72$$

Here we have $f^2(0) = 0100$, so $h(f^2(0)) = h(0100) = 3 \times 36 + 72 = 180$. This means that for $m = 3$ the above condition for $f^m(1)$ is fulfilled. Since $h(f^3(0)) = h(01000101) = 5 \times 36 + 3 \times 72 = 396 = 36$ is rational and unequal 0, the conditions of Theorem 10.1 are satisfied, from which we conclude that the turtle figure is finite.

As a picture, this is less spectacular than some other examples we saw before. Nevertheless, for me personally, it is of a special significance. At my farewell at Eindhoven University of Technology in September 2022, from colleagues I received a realization of this in stained glass, of more than half a meter in diameter. The ten diamonds around the center were made of colorless glass with a lot of relief. The ten quadrangles around them were made in colorless glass with hardly any relief. Finally, the twenty figures on the outside were made of twenty different types of glass in all the colors of the rainbow. This makes it a stunning object, in particular when light shines through.

Choosing other angles with $3h(0) + h(1) = 180$ give a finite turtle figure with $m = 3$ for the same reason. As an example, we give the turtle figure for $h(0) = 26$ and $h(1) = 102$:

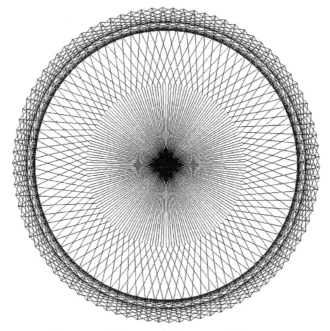

$$f(0) = 01, f(1) = 00, h(0) = 26, h(1) = 102$$

Higher values of m also provide examples. For instance, we have

$$f^3(0) = 01000101,$$

and that gives the requirement $5h(0) + 3h(1) = 180$ for $m = 4$. This applies, for example, to $h(0) = 18$ and $h(1) = 30$, giving the following turtle figure.

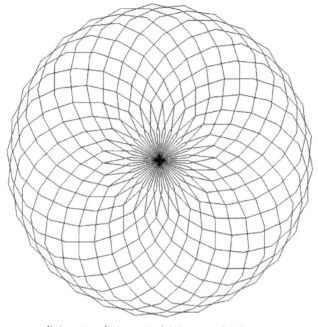

$$f(0) = 01, f(1) = 00, h(0) = 18, h(1) = 30$$

For even higher values of m, it is useful to make a table for the number of zeros and ones in $f^i(0)$ for several values of i. As usual, we write $|u|_0$ for the number of zeros in a word u, and $|u|_1$ for the number of ones in u. Because of $f(0) = 01$ and $f(1) = 00$ then $|f^{i+1}(0)|_1 = |f^i(0)|_0$ and $|f^{i+1}(0)|_0 = 2|f^i(0)|_1 + |f^i(0)|_0$, for every $i > 0$. This easily gives the following table:

m	$|f^i(0)|_0$	$|f^i(0)|_1$
1	1	1
2	3	1
3	5	3
4	11	5
5	21	11
6	43	21
7	85	43
8	171	85

Using this table shows, for example, that the above approach works for $h(0) = 120$ and $h(1) = 20$ and $m = 7$, namely $|f^{m-1}(0)|_0 = 43$ and $|f^{m-1}(0)|_1 = 21$, and

$$43 \times h(0) + 21 \times h(1) = 43 \times 120 + 21 \times 20$$
$$= 5160 + 420 = 5580 = 15 \times 360 + 180.$$

And indeed, this yields the following finite turtle figure:

$$f(0) = 01, \; f(1) = 00, \; h(0) = 120, \; h(1) = 20$$

A few chapters ago we gave a challenge to determine the finiteness of a particular turtle curve. It was the turtle curve for this sequence **p** and angles $h(0) = 140$ and $h(1) = -80$. Here our approach applies for $m = 8$. Using the numbers in the table for $m = 7$ yields

$$85 \times h(0) + 43 \times h(1) = 85 \times 140 - 43 \times 80 = 11900 - 3440 =$$

$$8460 = 23 \times 360 + 180.$$

The additional requirement that $171 \times h(0) + 85 \times h(1)$ is rational and unequal to 0 is easy to verify, so the turtle figure is indeed finite. More precisely, according to Theorem 10.1 it is finite and consists of at most $(k + k')n$ line segments, where k, k' are the lengths of $f^m(0)$ and $f^m(1)$, both being $2^8 = 256$. Moreover, we obtain $n = 18$ because all angles are multiples of $20°$, so multiplying by 18 yields a multiple of 360°. So according to Theorem 10.1 the number of line segments in the turtle figure is at most $(256 + 256)18 = 9216$. This is indeed <10,000, solving the challenge.

Other simple morphic sequences

We started this chapter by characterizing the simplest morphic sequences: purely morphic sequences $f^\infty(0)$ over only two symbols 0 and 1 where both $f(0)$ and $f(1)$ consist of at most two symbols. We've seen two of these: **t** and **p**. Are there any more? We are then mainly interested in non-trivial cases, in particular, sequences that are not ultimately periodic.

Also sequences of the shape $f^\infty(1)$ may be considered, but that won't produce new turtle figures. This holds since swapping zero and one, both in the definition of f and in the angles, will give exactly the same turtle figure. So we restrict to the form $f^\infty(0)$. Then to be well defined $f(0)$ must be of the form $0u$ for a non-empty word u. As the length of $f(0) = 0u$ is at most two, this is only possible if u consists of one symbol. If $f(0) = 00$ then $f^\infty(0) = 0^\infty$, and therefore periodic. That leaves only one possibility: $f(0) = 01$. Then we still have to define $f(1)$, and that must be a non-empty word of at most two symbols. This yields six cases: $f(1)$ can then be 0, 1, 00 , 01, 10 or 11. Let's take a closer look at each of these six cases.

If $f(0) = 01$ and $f(1) = 0$ then $\mathbf{f} = f^\infty(0)$ is called the *binary Fibonacci sequence*. We considered this already in an example on programming morphic sequences. In the next section, it will be considered in more detail.

If $f(0) = 01$ and $f(1) = 1$ or $f(1) = 11$ then $f^\infty(0) = 01^\infty$. That's an ultimately periodic sequence.

If $f(0) = 01$ and $f(1) = 01$ then $f^\infty(0) = (01)^\infty$, being periodic.

That leaves two more: if $f(1) = 10$ then we get the Thue-Morse sequence **t** which we already discussed extensively. In the remaining case, we have $f(1) = 00$, giving the period-doubling sequence **p** that we considered before.

Thus, this series of simplest morphic sequences only consists of a few ultimately periodic sequences and three others: **t, p** and the binary Fibonacci sequence **f**. So now let's focus on that binary Fibonacci sequence.

The binary Fibonacci sequence

Much better known is the regular *Fibonacci sequence*

$$0, 1, 1, 2, 3, 5, 8, 13, 21, 34, 55, 89, \ldots,$$

defined by $\text{fib}(0) = 0$, $\text{fib}(1) = 1$, and

$$\text{fib}(i + 2) = \text{fib}(i + 1) + \text{fib}(i) \quad \text{for all } i \geq 0.$$

So, except for the first two elements $0, 1$, every element in this Fibonacci sequence is the sum of its two predecessors. Many exciting properties are known about this Fibonacci sequence, but for that we refer to other sources like Wikipedia. We only note that this sequence was already described in the 1202 book *Liber abaci* by Leonardo of Pisa, which was later also called Fibonacci.

Clearly, the Fibonacci sequence is a sequence over the natural numbers, while the binary Fibonacci sequence

$$\mathbf{f} = f^{\infty}(0) = 01001010010010100101 \cdots$$

for $f(0) = 01$ and $f(1) = 0$ is a sequence over the symbols 0 and 1. At first glance, they seem quite unrelated. But let's investigate the numbers of zeros and ones in $f^n(0)$ for any value of n. For any word u over 0 and 1 the length $|u|$ of u satisfies $|u| = |u|_0 + |u|_1$. We easily create the following table:

n	$f^n(0)$	$|f^n(0)|_0$	$|f^n(0)|_1$	$|f^n(0)|$
1	01	1	1	2
2	010	2	1	3
3	01001	3	2	5
4	01001010	5	3	8
5	0100101001001	8	5	13
6	010010100100101001010	13	8	21
7	0100101001001010010100100101001001	21	13	34

For this initial part all these numbers $|f^n(0)|_0$, $|f^n(0)|_1$ and $|f^n(0)|$ are numbers from the Fibonacci sequence. This holds in general as we prove in the next theorem:

Theorem 13.2 *For every $n \geq 1$ we have $|f^n(0)|_0 = \text{fib}(n + 2)$, $|f^n(0)|_1 = \text{fib}(n + 1)$ and $|f^n(0)| = \text{fib}(n + 3)$.*

Proof: We prove this by induction. For $n = 1$ it is correct. Now we assume as the induction hypothesis that the theorem holds for n, and we prove that it also holds for $n + 1$, that is,

$$|f^{n+1}(0)|_0 = \text{fib}(n + 3), \quad |f^{n+1}(0)|_1 = \text{fib}(n + 2) \quad \text{and} \quad |f^{n+1}(0)| = \text{fib}(n + 4).$$

The number of ones $|f^{n+1}(0)|_1$ in $f^{n+1}(0) = f(f^n(0))$ is equal to $|f^n(0)|_0$, because every 0 in $f^n(0)$ is replaced by 01, yielding a 1, while every 1 in $f^n(0)$ is replaced by 0, and thus does not return any 1. So $|f^{n+1}(0)|_1 = |f^n(0)|_0$. Due to the induction hypothesis, we have $|f^n(0)|_0 = \text{fib}(n+2)$, so $|f^{n+1}(0)|_1 = \text{fib}(n+2)$, being one of the three claims we had to prove.

The number of zeros $|f^{n+1}(0)|_0$ in $f^{n+1}(0) = f(f^n(0))$ is equal to $|f^n(0)|_0 + |f^n(0)|_1$, because every 0 in $f^n(0)$ is replaced by 01, yielding a 0, while every 1 in $f^n(0)$ is replaced by 0, thus yielding a 0 as well. So $|f^{n+1}(0)|_0 = |f^n(0)|_0 + |f^n(0)|_1$. Because of the induction hypothesis, we have $|f^n(0)|_0 = \text{fib}(n+2)$ and $|f^n(0)|_1 = \text{fib}(n+1)$, so $|f^{n+1}(0)|_0 = \text{fib}(n+2) + \text{fib}(n+1) = \text{fib}(n+3)$, being the first statement we had to prove.

The remaining statement is $|f^{n+1}(0)| = \text{fib}(n+4)$, and this holds because of

$$|f^{n+1}(0)| = |f^{n+1}(0)|_0 + |f^{n+1}(0)|_1 = \text{fib}(n+3) + \text{fib}(n+2) = \text{fib}(n+4).$$
□

Turtle figures of the binary Fibonacci sequence

After the many turtle figures of other morphic sequences that we have seen, of course we also want to investigate turtle figures of the binary Fibonacci sequence **f**. We gave criteria for turtle figures to be finite or fractal. Is it possible to choose angles such that these criteria hold for the binary Fibonacci sequence? Unfortunately, this is still completely open. As for all sequences over 0 and 1 a finite figure is obtained by choosing $h(0)$ and $h(1)$ equal to each other and $\neq 0$, and a line is obtained if $h(0)$ and $h(1)$ are both equal to 0, and therefore fractal. Apart from these trivial instances I am not aware of any choices of angles for which the turtle figure of the binary Fibonacci sequence **f** is finite or fractal. But it is nice to see how some turtle figures of initial parts of **f** look like for various choices of angles. As is the case for most sequences, for many choices the resulting figure is just a big mess, but in other cases some nice patterns seem to appear. We now give a few examples:

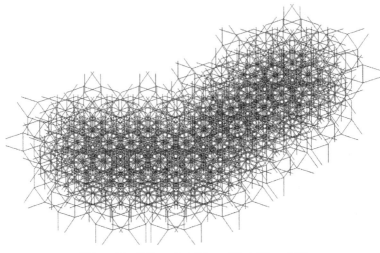

$$f(0) = 01, f(1) = 0, h(0) = 36, h(1) = 180$$

In this first example we chose the initial part of length 50,000, and the angles $h(0) = 36$ and $h(1) = 180$.

Choose the initial part of length 100,000, and the angles $h(0) = 0$ and $h(1) = 108$, which results in the following turtle figure:

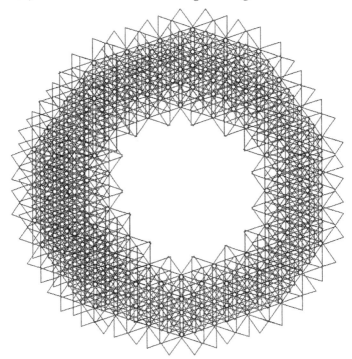

$$f(0) = 01, f(1) = 0, h(0) = 0, h(1) = 108$$

We give two more examples for angles that are multiples of 30°, and an initial part of length 100,000. To be precise: the turtle figure for the angles $h(0) = 180$ and $h(1) = 30$ and the turtle figure for the angles $h(0) = 60$ and $h(1) = -30$.

$$f(0) = 01, f(1) = 0, h(0) = 180, h(1) = 30$$

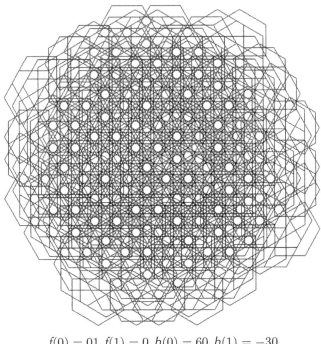

$$f(0) = 01, f(1) = 0, h(0) = 60, h(1) = -30$$

It is nice to look at these figures and look for patterns in them. Unfortunately, I am not aware of any theory explaining these patterns, so this ends our discussion on turtle figures of the binary Fibonacci sequence.

Frequency of symbols in morphic sequences

For a finite word u of length n we define the *frequency* of a symbol s as $\frac{|u|_s}{n}$. So this frequency is the fraction of symbols that are equal to s. If s does not occur in u it is 0, if u consists only of s it is 1, and otherwise it is something in between.

To define a similar notion of frequency for infinite sequences, a limit has to be taken, as we observed earlier in this book. It was defined as the limit of the frequencies of initial parts of the sequence of length n, for increasing n. This frequency of a symbol s in a sequence a is denoted by $[a]_s$. We already observed that for some sequences this limit does not exist at all. For ultimately periodic sequences we observed that it always exists, and is a rational number. Now we wonder how to investigate and compute frequencies of symbols in morphic sequences.

A first observation is that for morphic sequences this frequency does not always exist. Consider the morphic sequence $a = f^\infty(2)$ for $f(0) = 00, f(1) = 11, f(2) = 201$. We have

$$a = f^\infty(2) = 2010011041^408 1^8 0^{16} 1^{16} \cdots.$$

Except for the 2 at the beginning, this is exactly the example of a sequence a for which the frequency $[a]_0$ of the symbol 0 does not exist. This single 2 at the beginning has no effect on the frequency, by the same argument also for this variant the frequency $[a]_0$ of the symbol 0 does not exist. Similarly, $[a]_1$ does not exist. Only $[a]_2$ exists and is equal to 0 because the symbol 2 only appears at the beginning.

But how about other morphic sequences, like the three sequences **t**, **p**, and **f** playing the main role in this chapter?

The Thue-Morse sequence **t** was constructed in such a way that the zeros and ones were distributed as evenly as possible. So we expect $[t]_0$ and $[t]_1$ to both be $\frac{1}{2}$. Indeed that is what we will prove.

We restrict our analysis to purely morphic sequences $a = f^\infty(0)$ over two symbols 0 and 1. The same approach works for more general morphic sequences over more symbols, but then needs a slightly more complex notation. We also restrict to $[a]_0$: by using $[a]_0 + [a]_1 = 1$ the other frequency $[a]_1 = 1 - [a]_0$ is derived immediately from $[a]_0$. For finite words u we write $|u|$ for the length of u. Over the two symbols 0 and 1 we have $|u| = |u|_0 + |u|_1$.

Furthermore, we concentrate on the central idea and do not elaborate all arguments on limits in detail. The central idea is as follows: if $[a]_0$ exists then it is obtained as the limit of $\frac{|f^n(0)|_0}{|f^n(0)|}$ for increasing n. So for large n the value $\frac{|f^{n+1}(0)|_0}{|f^{n+1}(0)|}$ is approximately equal to $\frac{|f^n(0)|_0}{|f^n(0)|}$. In other words, for $u = f^n(0)$ for some large number n, the frequency $\frac{|f(u)|_0}{|f(u)|}$ is approximately equal to $\frac{|u|_0}{|u|}$. In the limit it will be exactly equal. More precisely, if $[a]_0$ exists and is equal to a number x, then for $u = f^n(0)$ for large n, in the limit we have

$$x = [a]_0 = \frac{|f(u)|_0}{|f(u)|} = \frac{|u|_0}{|u|}.$$

Now let's see how the shape of $f(0)$ and $f(1)$ may help to compute this value $x = [a]_0$. Write A for the number of zeros $|f(0)|_0$ in $f(0)$, B for the number of zeros $|f(1)|_0$ in $f(1)$, C for the number of ones $|f(0)|_1$ in $f(0)$ and D for the number of ones $|f(1)|_1$ in $f(1)$. The number of zeros $|f(u)|_0$ in the word $f(u)$ is then $A|u|_0 + B|u|_1$, because by applying f every 0 in u yields A zeros in

$f(u)$, and every 1 in u yields B ones in $f(u)$. Similarly we obtain $|f(u)|_1 = C|u|_0 + D|u|_1$. So we have

$$x = \frac{A|u|_0 + B|u|_1}{(A+C)|u|_0 + (B+D)|u|_1}.$$

Replacing every $|u|_0$ by $x|u|$ and every $|u|_1$ by $(1-x)|u|$, and removing common factors $|u|$ in the numerator and denominator, now gives

$$x = \frac{Ax + B(1-x)}{(A+C)x + (B+D)(1-x)}.$$

This yields $(A+C-B-D)x^2 + (2B+D-A)x - B = 0$. This gives the following theorem, for which we omit further details of the proof:

Theorem 13.3 Let $a = f^\infty(0)$ over two symbols 0 and 1, and $A = |f(0)|_0$, $B = |f(1)|_0$, $C = |f(0)|_1$, $D = |f(1)|_1$. Assume that exactly one value x exists such that $0 \le x \le 1$ and $(A+C-B-D)x^2 + (2B+D-A)x - B = 0$. Then $[a]_0$ exists and is equal to this value x.

In our example of a morphic sequence for which $[a]_0$ does not exist, there was also a 2 involved. Apart from that it would give $A = D = 2$ and $B = C = 0$, for which every x satisfies $(A+C-B-D)x^2 + (2B+D-A)x - B = 0$. So for this example, the unicity requirement of the theorem does not hold.

Now let's see what happens when choosing our examples $\mathbf{t}$, $\mathbf{p}$ and $\mathbf{f}$. For $\mathbf{t}$ we have $f(0) = 01$ and $f(1) = 10$, so $A = B = C = D = 1$. Then the equation is $0x^2 + 2x - 1 = 0$, indeed having exactly one solution $x = \frac{1}{2}$. So by Theorem 13.3 we conclude $[\mathbf{t}]_0 = \frac{1}{2}$, exactly what was expected.

For $\mathbf{p}$ we have $f(0) = 01$ and $f(1) = 00$, so $A = C = 1$, $B = 2$ and $D = 0$. Then the equation is $0x^2 + 3x - 2 = 0$ which has exactly a solution $x = \frac{2}{3}$. So by Theorem 13.3 we conclude $[\mathbf{p}]_0 = \frac{2}{3}$. This is slightly more surprising: while the lengths of $f^n(0)$ are all powers of 2, we end up in a frequency having 3 in the denominator.

The most surprising example is $\mathbf{f}$. There we have $f(0) = 01$ and $f(1) = 0$ and so $A = B = C = 1$ and $D = 0$. Then the equation is $x^2 + x - 1 = 0$. According to the abc formula we learned in school, there are two solutions:

$$\frac{-1 + \sqrt{5}}{2} \quad \text{and} \quad \frac{-1 - \sqrt{5}}{2}.$$

Among these two, only $\frac{-1+\sqrt{5}}{2} = 0.618033\cdots$ is in the required range between 0 and 1. So by Theorem 13.3 we conclude $[\mathbf{f}]_0 = \frac{-1+\sqrt{5}}{2}$. Surprisingly, this is *not* a rational number. The number $\phi = \frac{1+\sqrt{5}}{2} = 1.618033\cdots$ is usually called the *golden ratio*. So $[\mathbf{f}]_0 = \phi - 1 = \frac{1}{\phi}$, where the last equality follows from $x^2 + x - 1 = 0$.

By Theorem 13.2 we obtained that $|f^n(0)|_0 = \text{fib}(n+2)$ and $|f^n(0)| = \text{fib}(n+3)$ for every n. It follows that $[\mathbf{f}]_0$ is equal to the limit of $\frac{\text{fib}(n)}{\text{fib}(n+1)}$, for n going to infinity, which is known to be $\phi - 1 = \frac{1}{\phi}$ is. About Fibonacci numbers and their relationship to the golden ratio an amazing amount of remarkable facts is known, but these are outside the scope of this book.

We conclude this chapter by noting that for almost all morphic sequences for which we have seen turtle figures in this book, the frequency of 0 exists and is easily determined using Theorem 13.3. For example, the rosettes were based on $f^\infty(0)$ for f defined by $f(0) = 011$ and $f(1) = 0$. This gives $A = B = 1$, $C = 2$ and $D + 0$, yielding the equation $2x^2 + x - 1 = 0$. Writing this as $2(x+1)(x-\frac{1}{2}) = 0$ gives $x = -1$ or $x = \frac{1}{2}$. So by Theorem 13.3 we conclude $[f^\infty(0)]_0 = \frac{1}{2}$. Hence the frequency of 0 is exactly one half. So the numbers of zeros and ones are equally balanced, just like $\mathbf{t}$, while in this case it was not immediately expected from the definition of f.

In other examples of finite turtle figures, we often had $f(1) = 11$ or $f(1) = 111$. This gives $B = 0$ and $D = 2$ or $D = 3$ in Theorem 13.3, typically ending up with $[a]_0 = 0$. For example, $f(0) = 010$ and $f(1) = 11$ yields the values $A = D = 2$, $B = 0$ and $C = 1$, for which $(A + C - B - D)x^2 + (2B + D - A)x - B = 0$ results in $x^2 = 0$. According to Theorem 13.3 this yields $[a]_0 = 0$, begin quite a surprise when observing the sequence $a = f^\infty(0) = 01011010111101011010\cdots$.

Another example is the first fractal turtle figure we saw, where f was defined by $f(0) = 001111$ and $f(1) = 10$, yielding $A = 2$, $B = D = 1$ and $C = 4$. This yields the equation $4x^2 + x - 1 = 0$, for which the *abc* formula implies $x = \frac{-1+\sqrt{17}}{8} = 0.390\cdots$, plus another negative solution. So according to Theorem 13.3 we obtain $[f^\infty(0)]_0 = \frac{-1+\sqrt{17}}{8}$. Just like in the binary Fibonacci sequence, this is a number not being rational.

Challenge: frequency of 1%

Challenge:

Is it possible to create a morphic sequence $a = f^\infty(0)$ for which the frequency $[a]_0$ of zeros is exactly one percent, that is, $[a]_0 = \frac{1}{100}$?

14 | Looking back

In this last chapter, we look back at the main observations of this book.

Turtle figures of morphic sequences

Clearly, the main theme of this book was to make infinite sequences out of a few, most times only two, symbols, and then make pictures of them in an obvious way: the turtle figures. The simplest infinite sequences were periodic or eventually periodic. They gave nice, but not yet very exciting turtle figures.

Hence the next natural step was: how to make infinite sequences in a systematic way, still containing a lot of structure, but having more potential than just the ultimately periodic sequences. Here we arrived at *morphic sequences*. We do not claim this was the only possible way to proceed, but morphic sequences turned out to have suitable features: the definition is simple, a very simple program serves to generate the sequence, and a wide range of surprising turtle figures emerged. Although not widely known in mathematics and computer science, morphic sequences have been studied extensively. To give an idea: in 2003 the book *Automatic Sequences* by Jean-Paul Allouche and Jeffrey Shallit was published. This is a very thorough work of some 570 pages, many dozens of which only consist of references to other work. The main theme of this book is *automatic sequences*. Many equivalent characterizations are given, one of which is that a sequence is automatic if and only if it can be presented as a morphic sequence for a morphism f for which for all symbols s the words $f(s)$ have the same have length. For example, $\mathbf{t}$ with $f(0) = 01$ and $f(1) = 10$ and $\mathbf{p}$ with $f(0) = 01$ and $f(1) = 00$ are automatic, but $\mathbf{f}$ with $f(0) = 01$ and $f(1) = 0$ not automatic. The book

DOI: 10.1201/9781003466000-14

extensively presents the theory on morphic sequences. For instance, one theorem states that applying an arbitrary morphism to a morphic sequence, always results in either a morphic sequence or a finite word. The book allows slightly more general morphisms than we do: it also allows $f(s)$ to be the empty word for some symbols s. Due to this generalization, the result of a morphism applied to an infinite sequence may be a finite word.

We focused on giving turtle figures of morphic sequences. Before doing so, in the first few chapters, we presented some general theory, well-known in mathematics. In particular, we investigated several types of numbers, presented proof principles about them such as induction, and considered different kinds of infinity. Many of these observations play fives the following turtle figure in which for a role in observations we later made about infinite sequences and their turtle figures. For instance, it turned out that a number is rational if and only if the sequence of digits behind the decimal point is ultimately periodic, and the turtle figure of the sequence 0^∞ is finite if and only if the angle $h(0)$ is rational and not 0. These observations involve rational numbers, one of the types of numbers we considered. As an application of the theory on different kinds of infinity, we observe that from the countability of the set of morphic sequences, one easily concludes that infinite sequences exist that are not morphic.

This basic theory, the concepts of morphic sequences, and investigating turtle figures under the name *turtle graphics* were already known. To my knowledge, no previous results were known investigating properties of turtle figures of morphic sequences. I started working on this around 2014. Many of the results discussed in this book were already presented in my paper *Turtle graphics of morphic sequences*, which was published in 2016 in the international journal *Fractals*. Others' work most closely related to our work on turtle figures are the *recursive rosettes* of Gailiunas as discussed in Chapter 10. We saw how these can be presented as turtle figures of a very specific morphic sequence. However, this link with morphic sequences was not observed in the original work.

Among these turtle figures, we focused on two types: finite turtle figures, that is, figures that consist of only finitely many line segments, and fractal turtle figures having the wonderful property that a magnified copy of the set of all end points of involved line segments is contained in the set itself. For both types, we presented criteria from which that special property can be concluded. In the development of the theory the given proofs play a crucial role: thanks to such a proof we know for sure that the property is true, and by considering the proof in detail we know exactly the essential conditions in order to draw the desired conclusion. All the presented dozens of examples of turtle figures are constructed that way: define the morphic sequence and the corresponding angles in such a way that exactly these conditions are met.

Actually, I made many more turtle figures using my program than only the figures included in this book. For every criterion a game was: how to choose $f(0)$, $f(1)$, $h(0)$ and $h(1)$ such that all conditions are met? Several choices were entered in my program and the resulting turtle figure was inspected. Every time among a large number of figures a choice was made which one or two to include in the book. Here the figure should show what I wanted to show, but apart from that there was a preference for *beautiful* figures, whatever that may be. When do we like a figure? While a lot of issues in this book were described with mathematical precision, this is not possible for the notion of being *beautiful*. What is considered as beautiful may be quite personal. However, some general characteristics of *beautiful* figures may be given. In particular, there should be some balance between *order* and *chaos*. If an image is very regular, like the turtle figures of 0^∞ from Chapter 6 that look like stars, it is mostly *order*, and therefore boring. Such figures will not be nominated for the most beautiful figure of the book. At the other end of the spectrum, we saw turtle figures of random sequences created by a *random generator* in the first part of Chapter 8. They showed a mess without any structure: certainly no *order*, but plenty of *chaos*. Neither of these will be nominated for the most beautiful figure of the book. No, for that some of both will be preferred: some order and regularity, but also surprises and variations. It will have some balance between *order* and *chaos*. When choosing the figures to include in this book, this was an important criterion. But the choices could have been very different; they were mainly guided by my personal preferences.

Other types of turtle figures

In this book, the focus was mainly on finite and fractal turtle figures. For sure there are more types of turtle figures of interest, being worthwhile to study. As an example, we consider the following. By only allowing angles of multiples of $90°$, for all end points (x, y) of the drawn segments, the values x and y will be integers. Is it possible to make a turtle figure of a morphic sequence where for all integers x, y the point (x, y) occurs *exactly once* in the turtle figure? Or only integers x, y occur both being ≥ 0, again all occurring exactly once?

Indeed this is possible: for both questions, we give an example meeting the conditions.

First, let's go back to the very first morphic sequence we considered not being ultimately periodic. That was the *spiral sequence* spir obtained by

$$\text{spir}' = f^\infty(2) = 2101001000100001 \cdots$$

for $f(0) = 0$, $f(1) = 01$, $f(2) = 210$, from which spir was obtained by replacing the first and only symbol 2 by 1. We saw that for $h(0) = 0$ and $h(1) = h(2) = 90$ this produces a spiral as a turtle figure, that's why it was called the spiral sequence. Due to the right angles, the turtle figure only moves over points (x, y) for which x, y are integers. Due to the spiral shape, it never visits the same point twice. For the above question, there is one more requirement: are all (x, y) for which x, y are integers visited? Unfortunately, this is not the case: because an extra 0 is added after every 1, only roughly half of those points are reached. But that may be fixed by having some fewer zeros in the sequence. More precisely, we want the sequence

$$1110101001001000100010 \cdots$$

When choosing $h(0) = 0$ and $h(1) = 90$, the resulting turtle figure will be spiral-shaped again, but now meets the extra requirement for all integers x, y the point (x, y) is met exactly once. So it remains to present this sequence as a morphic sequence. This may be done by two extra symbols 2 and 3, and choosing $f(0) = 0$, $f(1) = 3$, $f(2) = 21$, $f(3) = 01$, resulting in

$$f^\infty(2) = 2130103001003000100030 \cdots .$$

This gives exactly the desired sequence by replacing every 2 and 3 by 1. Indeed, by choosing $h(0) = 0$ and $h(1) = h(2) = h(3) = 90$ the purely morphic sequence $f^\infty(2)$ gives the following turtle figure in which for all integers x, y the point (x, y) is met exactly once:

More exciting is the second question: here we only want to meet (x, y) for integers x, y with $x, y \geq 0$, and every such point (x, y) should be visited exactly once. We only give an example without going into the theory. We define

$$u = 2110332,$$

$$f(0) = 0u0, \ f(1) = 0u1, \ f(2) = 3u1, \ f(3) = 3h0,$$

$$h(0) = 0, \ h(1) = 90, \ h(2) = 0, \ h(3) = -90.$$

An initial part of the turtle figure of the purely morphic sequence $f^\infty(0)$ looks like this:

Here the turtle starts at the bottom left. It is nice to follow the turtle figure starting from that point and try to see the pattern by which ultimately all requirement points are visited exactly once.

More exciting pictures: cellular automata

A particular feature of all turtle figures of morphic sequences in this book is that they are obtained by executing a program of only a few lines. The programs for the various figures differ very little: only in the numbers and the definition of the morphism.

This raises a much more general question: how to obtain a nice figure as a result of running a program of only a few lines? This admits a wide range of possible answers, most of which are completely independent of turtle figures. Among all possible directions, here we will discuss only one: *cellular automata*.

In cellular automata, the starting point is a grid of dimension 1 or 2. Each grid point has a color, often black or white, and then there is a simple rule defining the color of any grid point in the next step based on the colors of its neighbors.

A famous example in dimension 2 is the *Game of Life*, dealing with only two colors. It was designed by the famous British mathematician John Conway in 1970. The rule is as follows:

> After a step, a grid point becomes black exactly if it is black and has two or three black neighbors, or if it is white and has exactly three black neighbors. In all other cases, the grid point turns white.

Here, the *neighbors* of a grid point are the eight grid points directly above, below, next to it, or diagonally above or below it. This rule looks quite arbitrary, but has a delicate balance between turning white or black. On this basis, all kinds of phenomena occur: some patterns remain stable, others return after two steps, or many more steps, for example 15 steps. Funny is a *glider*s: that is a pattern copying itself in a number of steps, in a shifted position. Even more funny is a *glider gun*: a pattern creating a glider after a number of steps. So, unlimited execution creates infinitely many gliders.

With the help of gliders, even basic logical operations may be simulated. Using these as building blocks an entire computer may be simulated as an instance of this game, based on this single very simple rule. This is an exciting and highly recommended topic; for more amazing issues on this Game of Life we refer to the internet.

Even in dimension 1, so only have a sequence of points in which every point has only two neighbors, already very surprising cellular automata may be composed. This is one of the themes extensively described by Stephen Wolfram in his 2002 book *A New Kind of Science*. As an example, we describe the following. We start in a long sequence of dots, the middle one being black, the rest white. Now the rule is as follows. Among three consecutive points p, q, r the middle one q turns black after one step if either exactly one of the three p, q, r is black, or if q and r are black and p is white. In other cases q becomes white. Converting this rule directly into a simple Lazarus program yields the following picture in which for every $n \leq 200$ the nth row shows what the sequence looks like after n steps:

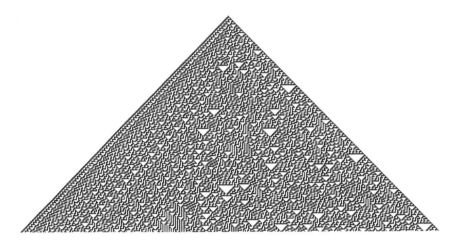

At the left, the picture looks quite regular, but at the right, all kinds of white triangles seem to appear quite arbitrarily, in all kinds of formats. Of course our rule is very solid and simple, so the behavior is not really arbitrary.

By slightly modifying the rule, and play around, a wide range of phenomena show up, ranging from order to chaos. In some sense, this looks quite similar to our turtle figures: a program only having a few lines of code, but generating exciting pictures showing a wide range of patterns.

Mathematical challenges

There is a wide range of mathematical problems that can be formulated briefly, and for which no heavy theory is required to understand the problem statement. Many of them are *open problems*: problems for which there is no known solution, even after extensive attempts by very smart people. We already mentioned Goldbach's conjecture: every even number > 2 can be written as the sum of two primes. Another one is Collatz conjecture. Here you start with any natural number $n > 1$. If n is even then n is divided by 2, otherwise n is replaced by $3n + 1$. This is repeated as long as $n > 1$. For example, starting with 7 yields 7, 22, 11, 34, 17, 52, 26, 13, 40, 20, 10, 5, 16, 8, 4, 2, 1. The conjecture is that for every starting number $n > 1$ the number 1 will be reached after finitely many steps. Of course, for small numbers, this is easily executed, and it is checked to hold for these numbers. But until now no one succeeded in proving that this holds for all starting numbers $n > 1$.

There are also problems for which a solution is known, but only a solution that involves a lot of heavy theory. A famous example is the last problem of Fermat: if $n > 2$ then there no integers $a, b, c > 0$ exist such that $a^n + b^n = c^n$. This problem was formulated around 1637 by Pierre de Fermat (1607–1665), even with the claim that he had a proof. However, that evidence was never recovered, and it is now suspected that it was wrong. This problem has been an open problem for centuries and a challenge to many mathematicians. Finally it was solved in 1993 by Andrew Wiles, by giving a very extensive proof exploiting very sophisticated theory, being far too hard for me to understand.

Apart from these open problems and problems only having very hard solutions, also various problems exist for having short and elementary solutions, although finding these solutions may be tough. We have encountered some of them as challenges in this book. Mathematical Olympiads at several levels, national and international, purely focus on this type of problems, for high school students. I myself had the privilege of participating in the International Mathematical Olympiad on behalf of the Netherlands in 1974 after being the winner of the Dutch Mathematical Olympiad. The problems for the International Mathematical Olympiad are often very difficult. Therefore they are of interest for anyone who likes mathematical challenges, independent of being a high school student or not. A notorious example is a problem from the 1988 International Mathematical Olympiad: prove that if

$$x = \frac{a^2 + b^2}{ab + 1}$$

for integers a, b, x, then x is a square. In the mean time quite elementary proofs for this claim are easily found on the internet. Checking whether such proof is correct is still feasible. However, finding such a proof yourself is really tough, in particular too hard for me.

Designing suitable problems for Mathematical Olympiads is a quite different kind of challenge. For me it was a great honor that 36 years after I was a participant myself, in 2010, a problem was selected for the International Mathematical Olympiad that was designed by me. In the next section we come back to it.

Some of the challenges in this book were in the spirit of Mathematical Olympiads. In particular, this was the case for the paint pot problem from Chapter 1, the monotone functions from Chapter 4, and the googolplex problem from Chapter 12. Among all challenges, these may be the most difficult.

These three problems all originated in my own research. They are so intriguing that I wanted to include them as challenges in this book, even though they are not directly related to the theme of this book. On the other hand, there were challenges directly connected to the theory just discussed. This applies to number theory from Chapters 2 and 3, and the sequences and turtle figures in Chapters 6, 8–10 and 13. Some other challenges described some game and the goal was to find out whether or not some particular final pattern was reachable. This holds for the marble box from Chapter 5, the knight moves from Chapter 7, and the very big number of challenges from Chapters 11 and 12. These are included not only because they are fun puzzles, but also because they deal with important techniques as invariants, or how to deal with extremely large numbers.

Finally, there is one more challenge in Chapter 14. That is the current chapter, and since the challenge is yet to come, it is not appropriate to classify it now.

Almost infinite

We already encountered the very large numbers googol and googolplex. We observed that these numbers are larger than anything one reasonably meets in nature. But it is possible to give simple definitions resulting in these kinds of numbers and even much larger numbers. A striking example is the *Ackermann function*, named after Wilhelm Ackermann (1896–1962). It is defined on pairs of natural numbers as follows:

$$
\begin{aligned}
A(0, n) &= n + 1, \\
A(m + 1, 0) &= A(m, 1), \\
A(m + 1, n + 1) &= A(m, A(m + 1, n)),
\end{aligned}
$$

for all $m, n \geq 0$. For every $m, n \geq 0$, exactly one of these three rules applies, a crucial ingredient for A being uniquely defined. At first glance, it does not look wild at all. However, trying to calculate $A(m, n)$ explicitly for small values of m, n soon results in very large numbers. The first results are quite modest: for every $n \geq 0$ we obtain $A(1, n) = n + 2$ and $A(2, n) = 2n + 3$. Choosing the first argument being 3 shows some more growth: for every $n \geq 0$ we have $A(3, n) = 2^{n+3} - 3$. But choosing the first argument being 4 or higher really explodes, for example, we obtain

$$A(4, 2) = A(3, A(4, 1)) = A(3, A(3, (A(4, 0)))) = A(3, A(3, (A(3, 1))))$$

$$= A(3, A(3, 13)) = A(3, 2^{16} - 3) = A(3, 65533) = 2^{65536} - 3,$$

being a number of more than 19,000 digits in the usual decimal notation. This is larger than googol, but still smaller than googolplex. But choosing higher arguments, for instance $A(4, 3)$ or $A(5, 2)$, values are obtained that are much larger than googolplex.

Speaking about such really big numbers, let's now consider the googolplex challenge. This challenge is about six boxes each containing zero or more marbles. Write $(n_1, n_2, n_3, n_4, n_5, n_6)$ for the situation where box number i contains n_i marbles, for $i = 1, 2, 3, 4, 5, 6$. The first rule allows to replace the values n_i, n_{i+1} by $n_i - 1$ and $n_{i+1} + 2$, for $i \leq 5$ and $n_i > 0$. The second rule allows to replace the values n_i, n_{i+1}, n_{i+2} by $n_i - 1$, n_{i+2} and n_{i+1}, for $i \leq 4$ and $n_i > 0$.

The first question is whether this may do infinitely many steps, so goes on forever. The answer to this is: no. Assume that it goes on forever. Looking at the first box, so at the value n_1, we see that by both rules it either remains equal (if $i > 1$), or decreases (if $i = 1$). Because we start by n_1 being a finite number, a number of marbles, this decrease may occur finitely often. So after a finite number of steps box 1 no longer participates, and n_1 will not change any more. But then the remaining infinite steps take place entirely in the last five boxes. By the same argument, there are infinitely many steps that take place in the last four boxes. And the same for three, and for two boxes. But it is clear that for two boxes it cannot go on forever: then only the first rule applies, and in each step, the number of marbles in the first box decreases. This gives a contradiction, based on the assumption that infinitely many steps are possible, so indeed it is not possible to do infinitely many steps.

The remaining part of the challenge asks whether it is possible to start in a situation in which only the first two boxes each contain one marble, and after finitely many steps reach a situation in which one of the boxes contains more than the gigantic number of googolplex marbles. At first glance, one may guess it will be impossible by only using these very simple rules. However, we will now show that it is possible. We start in $(1, 1, 0, 0, 0, 0)$, and applying the first rule a number of times gives

$$(1, 1, 0, 0, 0, 0), (0, 3, 0, 0, 0, 0), (0, 2, 2, 0, 0, 0),$$

$$(0, 1, 4, 0, 0, 0), (0, 0, 6, 0, 0, 0), (0, 0, 5, 2, 0, 0).$$

From here on, the first two zeros in the sequence, corresponding to the first two boxes being empty, do not participate any more and will be omitted in the notation. So now the goal is to get the gigantic number of marbles in one of the four boxes by using our rules, starting in $(5, 2, 0, 0)$. First we analyze more generally what kind of constructions are possible, starting by only three boxes. For every $k, n > 0$, starting from $(n, k, 0)$ applying k times the first rule gives $(n, 0, 2k)$, and then the second rule gives $(n - 1, 2k, 0)$. Applying this same series of steps n times in total, gives the following observation that we call $(*)$:

$(*)$ Starting in $(n, k, 0)$ it is possible to reach $(0, 2^n \times k, 0)$.

So powers of 2 may be constructed, already being large, but not yet large enough.

Now we consider four boxes, and investigate what may happen when starting in $(n, k, 0, 0)$, for $k, n > 0$. The first rule then yields $(n, k - 1, 2, 0)$. Then $(*)$ is applied, using $2 \times 2^{k-1} = 2^k$ this gives $(n, 0, 2^k, 0)$. Now applying the second rule results in $(n - 1, 2^k, 0, 0)$. So:

$(**)$ Starting in $(n, k, 0, 0)$ it is possible to reach $(n - 1, 2^k, 0, 0)$.

This new rule $(**)$ will be applied a number of times to our sequence $(5, 2, 0, 0)$. Here we use the following notation: $P(1) = 2$, and $P(k + 1) = 2^{P(k)}$ for $k \geq 0$. This is the key construction to obtain very large values: $P(2) = 2^2 = 4$, $P(3) = 2^4 = 16$ and $P(4) = 2^{16} = 65536$. The very large number $A(4, 2) = 2^{65536} - 3$ appearing in the Ackermann function satisfies $A(4, 2) = P(5) - 3$.

So $(5, 2, 0, 0) = (5, P(1), 0, 0)$. According to $(**)$ this goes to $(4, P(2), 0, 0)$. Applying $(**)$ some more times gives

$$(3, P(3), 0, 0), \quad (2, P(4), 0, 0), \quad (1, P(5), 0, 0), \quad (0, P(6), 0, 0).$$

So after doing these steps one of the boxes contains $P(6)$ marbles. We will now show that $P(6)$ is greater than googolplex, concluding the solution of the challenge.

Let's first estimate googolplex expressed in powers of two instead of powers of 10. Googol is 10^{100}, which is less than

$$16^{100} = (2^4)^{100} = 2^{400}.$$

Then googolplex is 10 to the power googol, being less then 16 to the power googol. According to the above observation this is less than 16 to the power 2^{400}, which may be written as

$$16^{2^{400}} = (2^4)^{2^{400}} = 2^{4 \times 2^{400}} = 2^{2^{402}}.$$

Here we follow the usual notation that $a^{b^c} = a^{(b^c)}$.

Now using $402 < 65,536$ gives

$$\text{googolplex} < 2^{2^{402}} < 2^{2^{65,536}} = 2^{2^{P(4)}} = 2^{P(5)} = P(6),$$

indeed showing that the box with $P(6)$ marbles contains more than googolplex marbles.

The impact of this solution goes beyond just solving a puzzle. This surprising result gives an example of a game with very simple rules, by which only finitely many steps are possible, but in which this number of steps can still be gigantic, even more than googolplex. Therefore this number of steps is much more than the number of atoms in the universe, or any other measurable quantity. This maximum number of steps turns out to be closely related to the Ackermann function. This issue originates from my own research, when I worked on techniques to prove termination of certain calculations, that is, show that these calculations cannot go on forever. When developing techniques to be very powerful, they should also work in situations where finishing these calculations may take a very long time. So in this research, it made sense to focus on these extremely long calculations, and from these calculations, this marble game was extracted.

In fact, this marble game is a variant of a problem that I submitted for the International Mathematical Olympiad in 2010. That problem was on the same six boxes with exactly the same rules. In the starting situation, there were six marbles, one in each box, and the aim was to show that it was possible to get exactly $2010^{2010^{2010}}$ marbles in the sixth box, while the other boxes were empty. Including some coding of the year in the problem was done before, and was easily done here. In fact this number is some larger than googolplex, but the key to solving it is exploiting the property $(**)$, exactly as we did above. Another complication is that the number had to be reached exactly in the last box while the other boxes were empty. This requires some extra analysis, but is not really difficult. By applying the second rule twice to the same position, the number in a box is decreased by two while the rest stays the same. The full problem may be solved by combining observations like these. Finally, this problem was indeed selected for the International Mathematical Olympiad in 2010. The selection committee made it even more difficult by not asking to show that it is possible, but by asking whether it is possible. As a consequence, many participants tried to show that it was not possible by trying invariance arguments, of course without success. Among problems for the International Mathematical Olympiad, this particular problem was considered to be quite hard and non-standard.

Challenge: the greatest value

Just like every chapter, this final chapter is concluded by a challenge.

In this final challenge, a process is executed in which integers A and B keep changing values. When it starts the number A is twice as large as B, we know $0 < B < 100$, and among the divisors of B (including 1 and B) there are exactly two prime numbers, exactly four squares (including 1), and exactly six other numbers.

Next the process continues with the instructions to be found just before the googolplex challenge.

Challenge:

What is the largest value of the number A during this process?

Index

Printed in the United States
by Baker & Taylor Publisher Services